W9-DBE-507

CATALOGUE

OF THE

NATURAL HISTORY DRAWINGS

COMMISSIONED BY

JOSEPH BANKS

ON THE

ENDEAVOUR VOYAGE 1768—1771

HELD IN THE BRITISH MUSEUM (NATURAL HISTORY)

PART 3: ZOOLOGY

Alwyne Wheeler

Bulletin of the British Museum (Natural History)
Historical Series Volume 13 (Complete)
London 1986

© British Museum (Natural History), 1986
ISSN 0068-2306
ISBN 0 565 09002-X

Published by British Museum (Natural History),
Cromwell Road, London SW7 5BD

British Library Cataloguing in Publication Data

Catalogue of the natural history drawings
commissioned by Joseph Banks on the Endeavour
voyage 1768–1771 held in the British Museum
(Natural History).—(British Museum (Natural
History) historical series; 13)
Pt. 3: Zoology
1. Natural history illustration—Catalogs
I. Wheeler, Alwyne
508 QH46.5

ISBN 0-565-09002X

Bull. Br. Mus. Nat. Hist. (hist. Ser.) 13: 1–172 (Complete)
Issued 31 July 1986

Typeset by Tradespools Ltd., Frome, Somerset and
Printed in Great Britain by Butler and Tanner Ltd., Frome, Somerset

CONTENTS

INTRODUCTION

This catalogue lists all the drawings of animals from James Cook's voyage on the *Endeavour* from 1768–1771 which were made by artists employed by Joseph Banks. Most of the drawings are the work of Sydney Parkinson (?1745–1771), but there are others by Alexander Buchan (?– 1769) and Herman Diedrich Spöring (?1733–1771).

The drawings were the property of Joseph Banks (1743–1820) and were kept in his London home in New Burlington Street from 1771–1777 and at 32 Soho Square from the summer of 1777 until his death in 1820. Under the terms of his will Banks's drawings, manuscripts, library, and collections were left to his curator–librarian Robert Brown (1773–1858) for life, with reversion to the Trustees of the British Museum on Brown's death, unless Brown and the Trustees had previously agreed to their transfer. The Trustees and Brown reached such an agreement and in 1827 the collections were transferred to the British Museum and Brown became Under Librarian for the Custody of the Banksian Collection.[1]

The Banksian Collection manuscripts and drawings were kept as a unit but the collection of books was placed in the Printed Book Department of the Museum. The zoological manuscripts of Daniel Solander and the drawings of animals were thus kept with the Banksian Collection of Plants. As late as August 1842 J.E. Gray, then Keeper of the Zoological Branch of the Natural History Department since 1840, referred to the animal drawings and manuscripts in the Banksian Collection of Plants (Gray, 1843*a*).

The manuscripts and drawings and those Banksian natural history books with annotations were transferred to the new building of the British Museum (Natural History) at South Kensington in 1880–1881. It is presumed that during this period the zoological manuscripts and drawings were separated from their botanical counterparts and became the responsibility of the Department of Zoology. Certainly the third volume of drawings was transferred from the Botanical Library to the Zoological Library on 8 March 1887 which suggests that the first two volumes were already in the latter library then.

The principal purpose of this catalogue is to record comprehensively the surviving animal drawings from the voyage of the *Endeavour*. However, the opportunity is taken to examine and discuss the history of the drawings, their arrangement and binding as well as giving short notes on the artists, none of whom survived the voyage. In addition, it has seemed relevant to discuss the manuscripts associated with the animal drawings, in particular Jonas Dryander's *Catalogue of the drawings of animals in the library of Sir Joseph Banks* and Daniel Solander's manuscripts relating to the voyage. A short discussion of the dispersal of the collection of animal specimens from the *Endeavour* voyage is preceded by some notes on the significance of the drawings to modern zoology.

General details of the voyage of the *Endeavour* are not given. These have been recounted many times, and are available in the introduction to a companion volume to the present paper in this journal (Diment *et al.*, 1984) where an extensive bibliography of the voyage, its principal participants, and its natural history are presented.

[1] British Museum, Trustees' Minutes 30 June 1827.

THE *ENDEAVOUR* ANIMAL DRAWINGS

The animal drawings are the work of three artists, Sydney Parkinson, Alexander Buchan, and Herman Diedrich Spöring. They are contained in three bound volumes; the binding believed to date from 1934. The drawings are mostly inlaid into larger sheets of 360 × 527 mm, but some are bound in directly on their original sheets (where these approximate, or exceed, the size of the volume). Because he used large paper for his drawings, these outsize sheets are mostly the work of Spöring. Several of these large drawings have been badly cropped by the binder. The drawings number 299 in total.

The drawings are currently (1985) in process of conservation and will be mounted as separate plates.

The artists have been studied in some depth by the late Averil Lysaght (1959, 1977, 1980) and brief notes derived from her later papers are given so that the contribution of each can be assessed more clearly.

Sydney Parkinson (?1745–1771), born in Edinburgh, was the son of Joel Parkinson, a brewer and a Quaker. He may have been trained as a draughtsman by William De la Cour, a Frenchman who established a School of Design in Edinburgh in 1760. Parkinson travelled to London in 1764 or 1765 and worked with James Lee, a nurseryman at Hammersmith, who engaged him to give drawing tuition to his daughter Ann. Within two years Parkinson was working for Joseph Banks drawing specimens, many of which were spirit preserved or stuffed skins, from Banks's earlier expedition to Newfoundland and Labrador in 1766 (Lysaght, 1971), and also numerous insects, chiefly Coleoptera and Lepidoptera, of tropical origin presumably in Banks's collection although possibly British Museum specimens. These drawings are still retained in the British Museum, Bloomsbury. Parkinson also copied some of the drawings prepared for Gideon Loten, a former Governor of Ceylon for the Dutch East India Company, and later resident in London. Some of these copies of drawings of Asiatic animals are preserved as a collection in the British Museum (Natural History); several were published by Pennant (1769) in his *Indian Zoology*. Recently, Parkinson has been the subject of a book edited by D.J. Carr, in which many of his zoological and botanical drawings from all sources are reproduced in colour, the zoological drawings being discussed there by Wheeler (1983). Parkinson died on 27 January 1771 when the *Endeavour* was homeward bound between Princes Island, West Java, and the Cape of Good Hope. Parkinson drew 268 of the drawings in the zoological collection.

Alexander Buchan (?–1769), was probably a member of the Berwickshire family. Little is known of his life despite considerable enquiry by Averil Lysaght. He joined Banks's team especially as a landscape artist and most of his work in this field is still preserved in the Department of Manuscripts at the British Library. However, he produced a number of water-colour drawings of animals on the first leg of the voyage, at Madeira and off Brazil. On Tierra del Fuego when Banks led his party on an expedition into the hills on 16 January 1769 Buchan suffered an epileptic fit from which he only partly recovered. He died on 17 April 1769 at Tahiti. Lysaght (1979) commented on the small quantity of his work after the *Endeavour* rounded Cape Horn, and also quoted Cook's journal recording his death in which Cook wrote 'he had long been subject to a disorder in his Bowels which are more than once brought him to the Very point of death and was at the same time subject to fits of one of which he was taken on Saturday morning, this brought on his former disorder which put a period to his life.' From this it is obvious that Buchan's health was precarious after the onset of his epilepsy at Tierra del Fuego, and the virtual absence of his drawings between Cape Horn and Tahiti needs no explanation. Besides which, this part of the voyage was across open sea, which certainly held no landscapes and from which limited zoological, and no botanical, material was available for drawing.

There are 21 drawings by Alexander Buchan in the three volumes of animal drawings (ff.120, 122, 135, 137, 171, 175–176, 192, 211–214, 216, 218, 221, 238, 291, 294–297). Thirteen of these are of invertebrates, the remainder of fishes. All were drawn at Madeira or off the coast of Brazil.

It is of interest to note that Dryander's *Catalogue of the drawings of animals in Banks's collection* lists a total of 32 Buchan drawings mostly of insects from localities (Madeira, Brazil, or Atlantic Ocean) visited by the *Endeavour*. In addition, a Buchan drawing of the starfish, Asterias tessellata Solander mss, from Rio de Janeiro is deleted in the list in pencil and Dryander has written 'destroyed' in the last column of the *Catalogue*. Two other drawings attributed to Buchan of the cockroach, *Blatta Germanica* L., have as their origin 'in nave', which having regard to the context (and the well-known affinity of cockroaches for ships) can only have been caught and drawn on the *Endeavour*. Therefore there were at the time of the compilation of Dryander's *Catalogue* a total of 35 Buchan drawings; this leaves 13 unaccounted for today.

Herman Diedrich Spöring (*ca* 1733–1771) was born in Åbo, then Swedish territory, now Turku in Finland, where his father was professor of medicine at the university. Educated at Åbo he left Sweden in 1755 and came to London where for eleven years he earned his living as a watchmaker. In 1766 he was employed as a clerk by Daniel Solander, then Assistant Keeper of the Natural History Department of the British Museum. His handwriting has been identified by Marshall (1978) in many of Solander's manuscripts in both botany and zoology. His role on the *Endeavour* voyage was that of amanuensis, transcribing the notes made by Solander and Banks as a fair copy, many of which survive for the botanical manuscripts. He also proved invaluable in repairing the ship's quadrant after it had been stolen by the natives on Tahiti. However, like Buchan and Parkinson, he was a versatile artist and numerous landscapes and ethnographic drawings in New Zealand and Tahiti were made by him. He was a highly competent zoological illustrator, working in pencil, and his drawings of rays, sharks, a bony fish, and crustaceans were not excelled in the zoological material.

His contribution to the illustrations of the *Endeavour* voyage had been largely overlooked until, with Averil Lysaght in the early 1950s, I recognized Spöring's name on one of his fish drawings, from whence it became simple to recognize his characteristic pencil drawings elsewhere. Lysaght (1979) recounts how she then rediscovered the 80 or so Spöring topographic and ethnographic drawings in several collections in the British Museum, all until then attributed to either Parkinson or Buchan.

Together with many of the crew of the *Endeavour*, and Banks's scientific team, Spöring contracted malaria and dysentery during the visit to Batavia. He died on 25 January 1771, two days before Sydney Parkinson, soon after leaving Princes Island, West Java, on the homeward-bound voyage.

There are a total of 9 drawings of animals by Spöring in the collection (ff.49, 50–52, 59–60, 99, 219–220). Dryander's *Catalogue* lists 9 Spöring drawings only; evidently none have been separated from the collection. Six of these are rays (including two of the sting-rays which gave their name to Stingrays Bay, later changed to Botany Bay), and sharks, one bony fish (f.99), and two crabs (f.219–220). The emphasis on large fishes might suggest that Spöring was especially capable of drawing large specimens (and indeed his renderings are more life-like and accurate than drawings of comparable subjects by either of the other artists) but his studies of crabs are so excellently executed that it is obvious that he was capable of producing accurate reproductions of small animals as well. Probably the explanation for his work on the rays and sharks was that Parkinson was simply overwhelmed by the quantity of botanical drawing required at this rich collecting ground, which caused the name change to Botany Bay, and Spöring was required to draw these impressive and unknown fishes before they were eaten or decayed.

The *Endeavour* animal drawings are arranged broadly in the order of the twelfth edition of Linnaeus's *Systema Naturae* (1766–1767). The names used in the arrangement are Solander's manuscript names which are written usually on the verso, sometimes on the recto, of each sheet. As Solander was very often employing Linnaean genera with their very broad concepts (often equivalent to families of fishes in modern ichthyology) the arrangement may at times appear to be quite arbitrary, although it was logical enough in the contemporary context. Problems occurred when Solander had introduced a new generic name, such as Nasutus in fishes, whose affinities he had not recognized, and which was merely bound in at the end of the volume containing bony fishes.

Another problem caused by the reliance on the *Systema Naturae* for arrangement was that the elasmobranchs (sharks and rays) and tetraodontiform fishes (trigger-fishes and puffer-fishes) formed the group Nantes of the Amphibia of Linnaeus (1766). As a result they are bound in volume 1 with the mammals, birds, and chelonians (Mammalia, Aves, Amphibia) to the surprise of some authors (e.g. Palmer, 1966, who apparently was confused by this).

All the drawings are annotated to some extent. Many bear short verbal accounts of coloration, or corrections to existing colour on the drawing. These are in the hand of Banks or Parkinson. Most of the drawings are named by Solander and the names are numbered in some way. It is evident that some listing of the drawings was undertaken probably as they were finished but it is difficult now to establish under what system the numbers were employed. The 5 land birds from Tahiti were numbered 1, 2, 3, 4, and 40 (ff.36, 34, 9, 35, and 8 respectively); Brazilian fishes are numbered 5, 7, 13, 14, 15, and 15 (ff.94, 122, 139, 135, 75, and 92 respectively) but it is difficult to see the logic behind these sequences. One probability is that the artists, either singly or collectively, were numbering their drawings in sequence, presumably keeping a master-list or lists, and for this reason there are extensive gaps in sequences (as in the Tahitian land birds). However, this does not account for the duplication in the Brazilian fishes. Most of the Madeiran animal drawings are numbered with the letter prefix T (it can sometimes be read as I), but again it proves difficult to reconstruct a logical sequence from the animal drawings alone. However, the use of a capital letter prefix to represent a geographical region is paralleled in Solander's method of note keeping (Wheeler, 1983) where notes on the fishes and the specimens were numbered serially with the prefix A from the Pacific Ocean, and B from New Zealand. Possibly a similar system was in use for the drawings but the notes, specimen numbers, and the drawing numbers are independent of one another and are not cross-referred.

The drawings are also labelled with the artist's name and in many cases the locality, both written in ink by Jonas Dryander. It seems most probable that Dryander wrote these annotations of artist during the period that he was compiling his *Catalogue of drawings of animals* (see below, p. 10) which dates them to the period 1772–1776.

It is probable that the present (1984) sequence of binding the drawings was adopted in Dryander's time. Indeed, it is logical to assume that Dryander, as Banks's librarian, would in a series of operations identify the artist for each drawing and the place at which it was made, both for his draft *Catalogue* and in order to label the drawings, and would then arrange the drawings in sequence, numbering them while doing so. As already noted, the sequence follows the twelfth edition of the *Systema Naturae* (1766–1767), although hitherto unrecognized genera (e.g. Nasutus, see above) or unidentified drawings were simply bound in at the end of the appropriate section. In the case of the fish genera involved, an attempt to place them in the correct systematic series was made in the final draft of Dryander's *Catalogue*, however, and this implies that the folio numbering, and thus sequence for binding, was earlier, perhaps pre-1772. (The numerals of the folios are written in pencil in large figures, but are not apparently in the hand of either Dryander or Solander.) There are other pencil numbers not part of the present sequence on some drawings. These may represent an earlier sequence but it is difficult to establish which it is as only a few drawings are so numbered. The main sequence of folio numbers was cited by authors as early as Heinreich Kuhl (1820), which shows that their use was firmly established before Banks's death.

There are two unsolved puzzles about the animal drawings from the *Endeavour* voyage. Firstly, they are not listed in Dryander's *Catalogus Bibliothecae historico-naturalis Josephi Banks* (Dryander, 1796:17), even though the Forster, Ellis, and Webber drawings were listed. Secondly, in his short history of the libraries in the British Museum (Natural History) Woodward (1904) listed, under Parkinson, within his list of Banks's library, the 'Original water-colour drawings of Plants and Animals made during Capt. J. Cook's first voyage . . .' eighteen botanical and one zoological volume. This was corrected by Sawyer (1971) in an otherwise identical entry to eighteen volumes of botanical and three of zoological drawings. Sawyer's list is clearly correct but why Woodward should have listed only one volume in 1904 remains a mystery; presumably it was a cataloguing omission arising from the recent transference of the present third volume from the Botanical Library to the Zoological Library (which took place on 8 March 1887). Possibly the first two volumes had not previously been catalogued but the recent transfer had been noted.

Few of the animal drawings by Sydney Parkinson are finished artwork, and some are merely pencil outlines often with notes on coloration which were to be finished later. The drawings by Spöring are detailed studies in pencil and are clearly finished drawings for record purposes. There seems to have been no intention to finish them in colour. Buchan's drawings are all finished water-colours. The unevenness in completion of the animal drawings has been touched on by Carter *et al.* (1981) who drew some general conclusions.

The presence of Buchan drawings of invertebrates and fishes from the first stages of the voyage can be interpreted that the experienced animal draughtsman Parkinson was advising or helping Buchan gain experience in this field, an artistic area far removed from the landscape studies he was engaged to practise. In addition, the approaches to Madeira were well-known to European sailors, and views of them would have been made by many earlier draughtsmen, so Buchan may also have taken the opportunity to draw animals as there was little other artistic employment for him. In the approaches to Rio de Janeiro he made a number of landscape drawings, even though this harbour was as well-known as Madeira, and there are others in the general area of Tierra del Fuego, but very few from the Pacific Ocean (Lysaght, 1980). Buchan's ill-health, and death at Tahiti, prevented him from making any considerable contribution to the artistic record of the voyage. His death moreover forced on Parkinson and Spöring a considerable burden of landscape drawing, a form which had not been designated as their first responsibility.

The animal drawings show most notably an emphasis on marine subjects, with a considerable number of oceanic, planktonic, or nektonic animals featured. This must be a result, not directly of the abundance of marine life, but to a lack of plant subjects while at sea, for Banks's interest in plants was greater than his interest in zoology. Nevertheless, the naturalists made critical and original observations on the nektonic animals such as crustaceans and their larvae, and plankton such as *Physalia physalis*. However, at landfalls Banks and Solander collected so many plants which they required Parkinson to draw that in many cases only single leaves, flowers, buds, or fruits were coloured on a drawing so as to give guidance for the completion of the finished drawing. Most of the animal drawings made in the Pacific Ocean were made near to or at landfalls, especially Tahiti, Australia, and New Zealand. A similar policy was adopted for these as for the botanical drawings, some colour being painted in, and colour notes made, but very few were ever finished. In Australia and New Zealand Spöring made most of his exact pencil drawings of animals, presumably while Parkinson was fully occupied with botanical drawing.

After the *Endeavour* returned to England it was Banks's decision to concentrate on the plans to publish the plant drawings. Many were redrawn by other artists and were later engraved on copper (Diment *et al.*, 1984) but the proposed grandiose publication was never achieved. By contrast the animal drawings were neglected, no attempt was made to produce finished or fully coloured artwork, and it seems that publication, although presumably envisaged as taking place after the plant volumes were published, never was a viable project.

As Carter *et al.* (1981) have already pointed out there was also a system of priority in the types of animals drawn. Fishes dominate the artistic record, birds, and marine invertebrates are also numerous, but mollusc shells and arthropods, such as insects, are scarcely represented. The choice of animal for drawing clearly depended on the likelihood of the preservation of its coloration and body form. The colours of fishes are highly fugitive, and after preservation in barrels of spirits of wine or rum (as they were on the *Endeavour*) would not have been discernable. Banks was already aware of drawings made from alcohol-preserved fishes after his Newfoundland and Labrador voyage, and even though Parkinson's drawings of them are accurate they are obviously made from dead fish, and they were not coloured. On his North American voyage Banks also had shot birds which were later stuffed and then drawn by Parkinson and Peter Paillou with only moderate success in the final presentation. The bias in the drawings towards soft-bodied and impermanently coloured animals such as coelenterates, salps, siphonophores, fishes, and birds, was thus a response to the naturalists' inability to preserve their colour and body form. The many molluscs, insects, and other arthropods which they are known to have collected were not drawn in any numbers, because they would have been prepared as dry specimens which would retain both their form and colour.

The bias in the animal drawing towards marine subjects can thus be seen as the result of conscious policy decision by Banks. The paucity of land animals represented in drawings and manuscripts is not because they have been lost or given away, as some authors have suggested, but is the obverse of the concentration on botanical studies while at the landfalls, and on animals only when plants were unobtainable.

DRYANDER'S CATALOGUE

Dryander's manuscript *Catalogue of the drawings of animals in the library of Sir Joseph Banks* is bound (rebound in 1947–1948) in half leather and comprises 251 numbered leaves, with 3 + 8 unnumbered original leaves at front and back; size 325 × 203 mm. A later entry on the title page reads 'J. Dryander's manuscript catalogue of the drawings of Animals in the

Library of Sir J. Banks arranged in systematic order'. The paper is watermarked C. Taylor, with a countermark of Britannia in an oval surround with a crown; it is not dated. Each leaf is ruled on both sides in reddish-brown ink, once horizontally near the head of the page, and with four vertical lines, from the horizontal line downwards, to give five unequal columns. These columns are used to give an indication of the medium of the drawing (see below), the scientific name of the animal, the locality or source of the specimen, and the artist concerned (Figure 1). The last column is usually left blank.

It represents a catalogue of all the animal drawings in Banks's collection arranged systematically under major group names, as current in the late eighteenth century, with each drawing listed under the genus name. Many drawings are identified by binominal name, with abbreviation of the name of the author of the binomen, and notes on the sex of the specimen in some cases. Most drawings have a locality of origin of the animal, or some other note, and all are attributed to various artists. Whitehead (1978) has listed these artists by name, in some cases identifying them with initials: the following list is reproduced from Whitehead with some modification, the most frequently used form of the name being in parentheses where this differs from the main entry: P. d'Auvergne (D'Auvergne); J. Backström (Backstroem); Barnes; Bolson; P. Brown; A. Buchan; J. Cleveley; N. Dance (Nath. Dance); T. Davies; G. Edwards; W. Ellis (William Ellis); Engleheart; G. Forster (Ge Forster); F. Frankland; S. Gilpin; J. Greenwood; W. King (Wilhelmina King); G. Metz (Gertrud Metz); J. Miller (Jas. Miller); J.F. Miller; U. Mole (Utrick Mole); F.P. Nodder; P. Paillou (Paillou); S. Parkinson; Chevalier Pinto; Roberts; J. van Rymsdyk; A. Schouman; J.E. de Sève (Seve); J. Sowerby; H. Spöring (Spo̊ring); J. Stuart; G. Stubbs; W. Watson (Dr. Watson); J. Webber; R. Wright.

Dryander adopted several shorthand symbols to abbreviate the information in the list. His first column details the medium employed for the drawing; Viz. × = 'Finished in Colors,' + = '[Finished] without Color', / = 'Sketch with Colors', — = [Sketch] without Colors, o = 'Copy upon transparent paper'. The third column in his catalogue is mainly a listing of the locality from which the subject came, but some other data are included. The localities are mostly abbreviations or contractions of an area name, but the nomenclature was, of course, eighteenth century usage. The localities are listed below; the abbreviation is copied exactly. Modern equivalents are given where needed.

Seven other abbreviations not referring directly to geographical locality or medium for the drawing were also adopted: a.v. Animal vivum (live animal); L.Y.A.M. Lady Anne Monson; O. Beng Original drawn at Bengal; Pen: Pennant (Thomas Pennant, 1726–1798, naturalist, author and correspondent of Banks); P.F. Pellis Farcta (stuffed skin); P. sal. Piscis salitus (fish preserved with salt); and s from an animal in spirit.

The list of localities and other abbreviations and symbols was evidently not compiled all at one time, for the writing differs (although all is by Dryander). Some of the entries are heavily inked, others are lightly inked, and there are two insertions into the alphabetical sequence. Looked at overall it is exactly the kind of list which would result from the first draft of a working catalogue. The main body of the catalogue, in which the drawings are listed, is very different. The writing is neat and consistent throughout the page and from page to page (with very few exceptions), the headings for class and order are always written centrally at the head of the page, the genus name is written in capitals at the head of the second column, and the entries for the drawings follow the same pattern exactly, except where information is lacking. Considered beside the introductory sheet of abbreviations the main listing is so consistent that it must represent a fair copy by Dryander from an earlier working list. The

Pisces

Thoracici

CHÆTODON

/	C. cornutus L. – – – – – –	Soc. Isl.	S. Parkinson	✓
/	C. lineatus L. – – – – – –	Soc. Isl.	S. Parkinson	✓
/	C. macrolepidotus L. – – –	Soc. Isl.	S. Parkinson	✓
/	C. saxatilis L. – – – – –	Soc. Isl.	S. Parkinson	
+	C. peregrinus Msp. – – – – – – –	P. sal.	J. F. Miller	
×	C. incisor Msp. – – – – – – –	Bras.	S. Parkinson	
×	C. Gigas Msp. – – – – –	Bras.	S. Parkinson	✓
×	C. cyprinaceus Msp. – – – – –	Oc.	S. Parkinson	✓
×	C. longirostris Brouss. – – – –	Sa. Isl.	W. Ellis	✓
×	C. triostegus Brouss. – – –	Sa. Isl.	W. Ellis	✓
/	C. stellatus Brouss. – – – – –	Hero. I.	W. Ellis	
/	——— – – – – –	Soc. Isl.	Ge. Forster	
/	C. glaucoparæus Brouss. – – –	Soc. Isl.	S. Parkinson	✓
/	——— – – – –	———	Ge. Forster	
/	C. Meleagris Brouss. – – – –	Fr. Isl.	Ge. Forster	
–	C. unicornis Brouss. – – – –	Soc. Isl.	S. Parkinson	✓
/	——— – – –	———	Ge. Forster	
×	——— – – – –	Sa. Isl.	J. Webber	
/	C. unicornis var. Brouss. – – –	Soc. Isl.	S. Parkinson	

Am. oc.	Americes littus occident
	(western coast of America)
Ascens	Island of Ascension
Bat:	Batavia India orientalis
Bras:	Brasilia
C.b.sp.	Caput bonae spei
	(Cape of Good Hope)
Chr. Isl.	Christmas Island
Fr. Isl:	Friendly Islands
Herv. I.	Hervey's Island
I.D.R.	Insula Diego Rays, Ind:or:
Kamt.	Kamtschatka
Kergu.	Kergulen's Land
M. Spitz	Mare prope Spitzbergen
M. pac:	Mare pacificum (Pacific Ocean)
M.P.B.	Mare pacificum boreali
	(northern Pacific Ocean)
Mad.	Madeira
Marque.	Marquesas
N. Cal.	Nova Caledonia
N.C.	Nova Cambria
	(New South Wales, Australia)
N.Z.	Nova Zelandia
Norf. I.	Norfolk Island
Ot.	Otaheite (Tahiti)
Oc.	Oceanus (open ocean)
Palm I.	Palmerston Island
Pr. Isl.	Princes Island, Ind. [ia]
Pul: Con:	or. [ientalis]
Rio Jan.	Pulo Condor Ind. or.
Sa. Isl.	Rio Janeiro
Soc. Isl.	Sandwich Islands
S. Geo.	(Hawaiian Islands)
Spizb.	Society Islands
St. L.	South Georgia
Sur.	Spizbergen
T. d. F.	Staten Land
Turtle I.	Surinam
Unal.	Terra del Fuego
	Turtle Island
	Unalashka

few exceptions where the handwriting is not consistent are clearly later insertions which the format adopted, with ample space between entries and one genus and another, was clearly designed to permit. Examples occur on f. 143 *Pleuronectes* Whiff Flounder . . . Dr Watson, and f. 17 *Mus* . . . N. America Col. Davies.

The date of compilation of the catalogue is difficult to establish but there are several indications of dating. Logically, a catalogue of Banks's collection of animal drawings might have been expected to be contemporaneous with Dryander's *Catalogus Bibliothecae* . . . (1796–1800). This great catalogue in four volumes included the drawings of animals from Cook's second and third voyages (vol. 2, p. 17) by George Forster, William Webber, and William Ellis, giving the number of folios in each case. In the case of the Forster collection, Dryander listed it as in two volumes totalling 261 folios, which agrees exactly with the present foliation in the two volumes (as some folios have two drawings mounted there are a

Fig. 1 Jones Dryander's manuscript *Catalogue of the drawings of animals in the library of Sir Joseph Banks* f. 145. The columns contain an indication of the medium of the drawing (see text for details), the scientific name of the animal represented, the locality or source of the specimen, and the artist. Note the number of *Chaetodon* species attributed to P. M. A. Broussonet, many are manuscript names which he communicated to J. F. Gmelin (1789).

total of 271 drawings). This demonstrates that this collection had been sorted, foliated, and one presumes mounted on sheets in some cases, before the production of Dryander's second volume (1796). The preparation of a catalogue of the drawings might therefore have been accomplished at the same time as the foliation, and certainly after it, and this catalogue might be dated to the late 1780s.

Although the omission has no bearing on the discussion of the date of preparation of Dryander's manuscript *Catalogue*, it must parenthetically be noted that the printed catalogue (Dryander, 1796–1800) does not list the original drawings of animals or plants from the *Endeavour* voyage, although the collected drawings from the second and third voyages are listed. It seems to be an inexplicable omission.

Jonas Dryander (1748–1810) became Banks's librarian after the death of Daniel Solander in 1782. The dates of his working for Banks thus provide a narrowed period of time for the compilation of the catalogue of drawings.

Within the catalogue there is some evidence for dating from the scientific nomenclature used. Thus, on f.65 the genus *Aptenodytes* is entered in the normal manner and this name was published by Forster (1781*a*). On f.43 and f.75 the bird genera *Callacas* and *Chionis* are both entered in pencil, although the entries for the appropriate drawings are given in the normal manner in ink. These generic names were published by Forster (1788). On f.175 two entries under the genus *Clupea* refer to P.M.A. Broussonet's *Ichthyologia*, which was published in 1782. Both species are entered in ink in the normal manner and are contemporaneous with the major part of the catalogue. From internal evidence of the nomenclature used it seems that the main catalogue was compiled after 1781–1782 and before 1788.

Confirmation of the earlier date can be found in the reference on f.11 to Forster's description of *Felis capensis* (Forster, 1781*b*), the entry written contemporary with the main catalogue. Many names attributed to Broussonet particularly within the genera *Chaetodon* (ff.145–147), *Sciaena* (ff.155–157), and the manuscript name Meandrites (f.157) are often written in pencil (the *Chaetodon* entries are in ink). These were names given to the fishes represented in the drawings, and often written on the drawing, by Broussonet. His published *Ichthyologia* (Broussonet, 1782) was sub-titled 'Decas I' but no further publication appeared under this title. These manuscript names, which must date from the period 1780–1782 or 1786, when he was in London and worked on Banks's collections, were presumably intended to be published in later decades of the work, and are thus treated somewhat tentatively by Dryander by entering some in pencil. Some of the *Chaetodon* names were published by Gmelin (1789) from a personal communication by Broussonet.

These varied indications for dating the Dryander manuscript give a broad range of the 1780s, narrowing down on single pieces of evidence to between 1782 and 1788. It therefore seems possible that Dryander compiled this catalogue around 1785.

The systematic arrangement of the catalogue basically follows the arrangement of Linnaeus's (1766–1767) twelfth edition of the *Systema Naturae*. There are, however, additions of genera which were not represented in the published work. With the case of the well-known, but wrongly-sited Kanguru, Dryander followed Solander's placement in Mammalia – Glires, he followed Solander for the new genus Nectris (between *Procellaria* and *Diomedea* in Aves-Anseres), and in both cases would have been influenced by the manuscript notes and even the sequence of binding. However, with new taxa from George Forster's drawings, Dryander, having no manuscripts for guidance, presumably had either to follow the sequence of the binding or take decisions as to the correct order to adopt. It is interesting that the Dryander *Catalogue* in some places improves on the sequence of arrangement adopted in the drawings. In the *Endeavour* drawings collection the sequence of arrangement is also based on the twelfth edition of the *Systema Naturae*, but at the end of the volume containing fishes there are six drawings (ff.205–210) which are out of sequence. The reason for this is quickly evident; two of them (ff.209, 210) are unnamed, except for vernacular names, and the remainder represent genera which were novel and thus not placed in the system. These were Nasutus (Solander ms) – a synonym of *Gomphosus* Lacepède, and Dentex (Solander ms, non Cuvier) – the species of which are properly synonyms of *Saurida* Cuvier & Valenciennes and *Synodus* Scopoli. In Dryander's catalogue, however, these two genera had been placed in Pisces – Thoracici, after *Labrus*, and Pisces – Abdominales, after *Salmo*, thus correctly placing them in juxtaposition to their nearest relatives in the context of eighteenth century ichthyology. Possibly Dryander was sufficiently good an ichthyologist to have made these critical determinations, but this seems unlikely. It is surely more probable that Broussonet, well known for his studies on fishes, re-examined these new genera of Solander's, for which drawings, manuscript accounts, and in the case of Nasutus at least specimens (still preserved in the British Museum (Natural History)) were available, and correctly allied them with their relatives in the Linnaean system. Whether this was done during his visits in Solander's life-time (1780–1782) or on his later visit in 1786, after Solander's death, is not known. The latter seems more probable.

There are two final points about Dryander's catalogue. Firstly, the detailed listing by Dryander makes it possible to confirm that all the drawings listed in the mid-1780s are still present. This shows that in the class Insecta a considerable number of *Endeavour* drawings, especially those made by Alexander Buchan are no longer in the main collection. (These are discussed in more detail earlier, see p.7.) Secondly, most of the entries for the Cook voyage drawings are obliquely crossed through in pencil. In many cases it is a strong pencil stroke running through the set of entries, sometimes as many as twelve crossed through at once; in others one or two drawings specifically out of a set are scored through. This system of cancelling the entries looks crude, even careless, but is in practice most specific and exact. Its significance cannot be explained. It seems unlikely to have been done as a result of checking the individual drawings, as these were in four separate collections, for then an individual mark would have been more appropriate (indeed some, notably the Ellis bird drawings, are individually marked with a tick in possibly the same coarse pencil). Possibly a separate list of Cook voyages drawings was compiled from this catalogue, and the entries in the original were then cancelled. If this was so then the later list seems not to have survived.

SOLANDER'S ZOOLOGICAL MANUSCRIPTS

The surviving manuscripts of Daniel Solander in British collections have recently been catalogued by Diment & Wheeler (1984). Those that relate to the *Endeavour* voyage are discussed below in greater detail, but are referred to the Diment & Wheeler catalogue by the item numbers; these numbers (e.g. D. & W. 40c) are also used in the following catalogue of the animal drawings to indicate each manuscript, although in addition abbreviated titles (e.g. P.A.O.P.) are also quoted. The abbreviated titles are identified below.

ORIGINAL MANUSCRIPTS

Original descriptions of fishes and other animals obtained on Cook's First Voyage, with notes from the Iceland Voyage. 382 p.; 20 cm. (D. & W. 40)

This is a bound volume containing five manuscripts, the first four of which pertain to the *Endeavour* voyage. Each relates to a different geographical area and is compiled in the order of the acquisition of new specimens (or, at least, in the order in which they were described although the chronological sequence was ultimately fundamental). The manuscript is in the hand of Solander although indexes to two sections were made by Spöring. It is mostly written in ink but there are pencil notes and several smaller slips of paper are inserted although these are mostly vocabularies or rough notes of little direct relevance to the main manuscript.

This manuscript was held in the Botany Department of the Museum until 1875 when it was transferred to the Library of the Zoology Department.

Each section of the manuscript is foliated separately in a contemporary hand; the whole was recently (1980) paginated to include all slips and covers. These numbers are referred to as f. (= folios) and p. (= pages) respectively.

The four *Endeavour* manuscripts are:

Pisces Australiae 54 numbered folios, total pagination 1–76; (D. & W. 40a), referred to here as P.A. Australia in this context refers to New Zealand. Contains descriptions of 41 species of fishes, and one bird, *Pelecanus leucogaster* (p.59) on a tipped-in, pencil-written slip. An alphabetically arranged index to the manuscript, which also serves as a list of specimens preserved (69 fishes in total) and the serial number allotted to each species, occupies p.62–75.

Pisces &c. Novae Hollandiae 19 numbered folios, total pagination 77–106; (D. & W. 40b), referred to here as P.N.H. Contains descriptions of fishes (13 species), one bird, *Falco vidua* (p.79), and a crustacean, *Cancer lituratus* (p.88). The running head to the pages identifies the group entered on the page, as Pisces or Pisces & Insecta (Insecta Aptera for *Cancer*), and Amphibia (these last referring to the rays and sharks which were placed in Amphibia Nantes). There is no index to this section. Nova Hollandia (New Holland) was the name in use for Australia at the time of the voyage.

Pisces & Anim. caetera Oceani Pacifici 140 + 7 numbered folios (numerous unfoliated leaves); p.107–298 (D. & W. 40c) referred to here as P.A.O.P. This manuscript comprises several parts. Descriptions of animals from Tahiti occupy f.1–128 (p.113–248), from f.129–140 (p.249–260) fishes only are described 'got at the other Islands in the South Seas' (quoted from a note on the inside front cover), and separately foliated 1–7 (p.261–267) are descriptions of birds from Tahiti and Raiatea. An extensive index to the manuscript (p.285–292) lists the species as they are described, also giving the serial number of the specimen, the number of specimens preserved, the number of the 'cagg' (a small barrel) in which they are preserved; vernacular names are also given here as well as in the formal description for many of the fishes.

In this manuscript Tahiti was originally named George Land, or Otaheite and variant spellings. Most of the fishes described from 'other Islands in the South Seas' were captured at Ulhaietea (= Raiatea) p.249–254, but a tuna, identified as *Scomber thynnus*, Linnaeus, 1758 was described from Ohitirhoa (= 21°47′ 151°9′) on August 13, 1769 (p.255–259).

This manuscript is mostly concerned with fishes, but some birds are described, as noted above, namely *Ardea nigricans*, *Columba pectoralis* (p.261), *Hirundo fuliginosa*, *Anas fasciata* (p.262), *Ardea nivea*, *Cuculus otaheitensis* (p.263), *Pelecanus otaheitensis* (p.265), *Alcedo superhitiosa* (p.266), and *Sterna fuliginosa* from Ulhaietea (p.267). Animals other than fishes and birds are described within the body of the text on Tahiti as follows, *Lacerta soleata* (p.239), *Anguis platura* (p.221), *Asterias crasissima* (p.241), *Cancer escarlatinus* (p.198), *Cancer fasciatus* (p.188), *Cancer marmoratus* (p.197), and *Sepia octopodia* (p.159).

There are numerous sheets bearing pencil notes, some being descriptions of animals, others being words collected for vocabularies of the Tahitian language.

The entries on pp.255–259 have been deleted by means of vertical black and red lines running the length of the page. These pages were not transcribed into the fair copy of the manuscript (see D. & W. 41) but are present in C.S.D. (D. & W. 42), see below.

Animalia Javanensia & Capensia 30 numbered folios, total pagination 299–352; (D. & W. 40d) referred to here as A.J.C. This short manuscript is divided into two, *Animalia Javanensia* (p.301–304) and *Animalia Capensia* (p.307–330). There are several sheets of pencil notes, some of which may have no relevance to these sections of the manuscript, and p.345–347 are descriptions of birds (*Anas circia*, *Charadrius pluvialis*, and *Falco ossifragus*) some of which refer to the Iceland journey and are related to the fifth manuscript *Pisces Islandici*, which despite its title refers to mammals and birds in Iceland as well as fishes.

The first entry for this manuscript is dated 8 October 1770 and briefly described a bat *Vespertilio vampyrus* and a plover, *Charadrius pluvialis*, one of four shot on a small island close to Pulo Pari (5°52′S. 106°38′E) in the Agenieten group (Groves, 1962). Two other Javan animals are described, *Sciurus musarum* and *Cervus plicatus* (p.301), and there is a list (p.303) of vernacular names of animals from Princes Island (= Prinsen-eiland, off West Java) including mammals, birds, a turtle, and marine invertebrates.

The *Animalia Capensia* includes descriptions of nine birds, *Rallus cristatus*, *Diomedea demersa*, *Scolopax leucocephala*, *Anas leucops*, *Anas maculatus*, *Anas pilearis*, *Anas monstrosa*, *Ardea pelearis* (p.307–314), *Otis pavoninus*, *Vultur protheus* (p.322–325), and several mammals, *Capra torticornis*, *Capra migratorius*, *Capra spiricornis*, *Capra rupestris*, *Bos equinus/barbatus*, *Simia ursina* (p.315–322) and *Viverra suricatt* (p.327). By the inclusion of a reference to a drawing (Fig. Pict.) it is obvious that Banks had purchased drawings from a local source in South Africa; (Wheeler (1984*a*) suggested the Brants, who entertained Banks when the *Endeavour* was at the Cape). It is possible that some of the animals described were seen in captivity at the Cape.

There is a striking contrast between the highly organized and neat note making of the first part of the voyage and the rather disordered notes in this section of the manuscript when Solander and most of Banks's team were sick and Spöring and Parkinson dead, after their stay in Batavia.

Slip catalogue containing descriptions of animals in the British Museum and other collections, including species collected by Solander, some during Cook's first voyage 27 volumes. 4842 sheets. Mostly 10 × 16 cm but varies. Referred to here as S.C. (D. & W. 45)

This *Slip Catalogue* was originally kept as loose sheets in Solander boxes but the sheets are now bound in volumes.

The *Slip Catalogue* was described by Diment & Wheeler (1984) and was discussed by Wheeler (1984*a* & *b*) who concluded that it was essentially a loose-leaf filing system devised by Solander to keep his zoological notes in an adaptable and readily available form. Some of the notes date from the period before he left Sweden in 1760, others record his own collections in England, a few record descriptions of *Endeavour* specimens, while many record notes taken while cataloguing the collections of the British Museum, and private collections like those of Joseph Banks, Lady Anne Monson, the Duchess of Portland, and Lady Bute. The Mollusca volumes are rich in entries referring to the Portland collection.

Further detailed study of the *Slip Catalogue* would be a valuable exercise because it presents a partial survey of the holdings of these important collections and illustrates the standing of collections in London in the middle years of the eighteenth century, many of which were later dispersed and the specimens have disappeared or lost their identity. Wilkins (1955) made a brief analysis of the Mollusca volumes. For the present, however, only descriptions of animals which come from localities visited by the *Endeavour*, or for which there is other literary or artistic evidence of its origin on Cook's first voyage, are listed. These are given by volume, folio number, Solander's name (which is on many occasions unpublished), and notes of the locality.

MAMMALIA – f.9 *Simia satyrus* (Batavia, 13 December 1770), f.90 *Kanguru saliens* (New Holland), f.101 *Cervus axis* (Java).

AVES – f.23 *Psittacus* (New Holland, possibly not *Endeavour* specimen), f.126 *Anas antarctica* (Tierra del Fuego), f.151 *Diomedea exulans* (23 December 1768, 3 March 1769), f.157 *D. exulans* var. (3 February 1769), f.159 *D. exulans* var. (2 October 1769, 6 January 1770, 11 April 1770), f.160 *D. antarctica* (1 February 1769), f.162 *D. profusa* (3 February 1769, 15 February 1769), f.164 *D. impavida* (11 April 1770), f.168 *Pelecanus aquilus* (America meridionali), f.170 *P. antarcticus* (Tierra del Fuego), f.171 *P. sectator* (24 December 1769), f.174 (New Holland), f.180 *Phaeton athereus* (Tahiti), f.181 *P. erubescens* (southern ocean, Tahiti), f.190 *Larus gregarius* (Tierra del Fuego), f.194 *L. crepidatus* (within the tropics), f.196 *L. fuliginosus* (Rio de Janeiro), f.197 *L. nigricans* (Brasil), f.201 *L. skua* (Ocean Australiam), f.208 *Sterna nasuta* (New Holland), f.210 *S. nigripes* (Tahiti), f.240 *Otis pileata* (Bustard Bay, New Holland), f.267 *Loxia nitens* (Brasil), f.275 *Motacilla avida* (28 September 1768), f.277 *M. velificans* (3 September 1768).

AMPHIBIA – f.14 *Testudo mydas* (New Holland), f.16 *T. caretta* (south Atlantic Ocean), f.104 *Boa pelagica* (Mare Pacifico), f.139 *Anguis marina* (New Holland – New Guinea), f.143 *Raja areata* (Totaranui, New Zealand), f.153 *R. nasuta* (Totaranui, N.Z.), f.159 *R. aquila* (Murderer's Bay, N.Z.), f.162 *R. rostrata* (New Holland), f.189 *Squalus lima* (off Novam Zelandica), f.193 *S. mystax* (New Holland, Sting Ray's Bay), f.207 *S. carcharias*, f.211 *S. glaucus* (Osnabrugh Island), f.238 *Balistes monoceros* (Atlantic Ocean), f.261 *Diodon erinaceus* (Atlantic Ocean, 7 October 1768), f.296 *Syngnathus pelagicus* ("Fuco natante Oceani Atlanti").

PISCES – volume 1, f.8 *Muraena guttata* (Madeira & Rio de Janeiro), f.99 *Coryphaena hippurus* (meristics for two fishes given); f.107 *C. novacula* (Madeira), f.130 *Scorpaena patriarcha* (Madeira), f.134 *S. chorista* (Madeira), f.156 *Pleuronectes rhomboides* (Madeira), f.161 *Chaetodon gigas* (Brasil at Rio de Janeiro), f.166 *C. cyprinaceus* (mid-Atlantic, 15 October 1768), f.168 *C. incisor* (Brasil), f.176 *C. luridus* (Madeira), f.193 *Sparus sargus* (Madeira), f.200 *S. griseus* (Madeira), f.208 *S. mundus* (Madeira), f.216 *Callyodon rubinosum* (Madeira).

– volume 2, f.11 *Labrus lunaris* (Madeira), f.27 *Sciaena angustata* (Madeira), f.30 *S. labiata* (Brasil), f.34 *S. rubens* (Brasil), f.57 *Perca asellina* (Rio de Janeiro), f.59 *P. nebulosa* (Brasil), f.63 *P.*

decorata (Madeira), f.67 *P. imperator* (Madeira), f.94 *Scomber scombrus* (Madeira), f.95 *S. pelamis* (Rio de Janeiro), f.101 *S. thynnus* (Ohitirhoa, Pacific Ocean), f.106 *S. lanceolatus* (Thrum Cap Island, Pacific Ocean), f.111 *S. serpens* (Canary Islands), f.115 *S. falcatus* (Brasil), f.119 *S. trachurus* (Madeira), f.121 *S. amia* (Brasil), f.125 *S. saltatrix* (off Brasil), f.128 *Mullus barbatus* (Otaheite), f.209 *Mugil albula* (Otaheite), f.213 *Exocoetus volitans* (Atlantic Ocean), f.269–272 are lists of fishes from Brasil and Rio de Janeiro, f.274–279 are lists of fishes from Madeira with vernacular names.

MOLLUSCA – volume 1, f.13 *Fasciola pelami* (in *Scomber pelamis* Atlantic Ocean, 1 October 1768), f.14 *Fasciola tenacissima* (in *Squalus glaucus* southern ocean 11 April 1769), f.17 *Sipunculus piscium* (in *Scomber pelamis* 1 October 1768), f.19 *Limax ramentaceus* (southern ocean 1, 2 October 1769; 11 January 1770), f.23 *Mimus volutator* (Atlantic Ocean, 4 October 1768, southern ocean 13 March 1769, 11 April 1770), f.26 *Doris complanata* (southern ocean 19 September 1769, 13 April 1770), f.44 *Actinia natans* (southern ocean, 12 April 1770), f.50 *Dagysa gemma* (numerous localities), f.52 *D. nobilis* (no data); f.53 *D. saccata* (Atlantic Ocean near Spain, 3 September 1768), f.55 *D. volva* (Atlantic Ocean, 3 October 1768), f.57 *D. limpida* (Atlantic Ocean, 4 October 1768), f.58 *D. lobata* (Atlantic Ocean, 4 September 1768), f.60 *D. corputa* (Atlantic Ocean, 2 September, 6 September 1768, 6 October 1769), f.62 *D. vitrea* (Atlantic Ocean, 7 October 1768, southern ocean, 3 February 1769, 13 April 1770), f.64 *D. vitrea* (no data), f.66 *D. rostrata* (Atlantic Ocean, 1768, southern ocean, 2 October 1769), f.68 *D. strumosa* (Atlantic Ocean near Straits of Gibraltar, and off New Holland, 23 April 1770), f.70 *D. serena* (southern ocean, 2 October 1769, 11 January 1770), f.72 *D. polyedra* (southern ocean, 2 October 1769), f.80 *Holothuria physalis* (Atlantic ocean), f.83 *H. physalis* (surface between the Tropics, 7°S Lat.), f.84 *H. physalis* (Atlantic Ocean, 22, 23 December 1768), f.86 *H. obtusa* (Pacific Ocean, 3 February 1769, 11 January, 11 April, 23 April 1770), f.89 *Scyllaea pelagica* (surface of Atlantic in floating algae), f.97 *Sepia octopodia* (Madeiran vernacular), f.99 *Calliroe bivia* (surface of tropical Atlantic), f.102 *Medusa rutilans* (Atlantic Ocean between the Tropics), f.104 *M. porpita* (Atlantic between Madeira and the Canaries; southern ocean, 13 April 1770), f.107 *M. punctulata* (Rio de Janeiro), f.110 *M. plicata* (between Tierra del Fuego and Staten Land), f.112 *M. radiata* (off Rio de Janeiro, 13 April 1770 (*sic*); New Holland, 23 April 1770), f.113 *M. fimbriata* (Rio de Janiero harbour), f.114 *M. vitrea* (southern ocean, 19 September 1769, 2 October 1769), f.116 *M. crystallina* (off Brasil), f.117 *M. limpidissima* (Tierra del Fuego), f.119 *M. obliquata* (near Tierra del Fuego), f.120 *M. pellucens* (off Brasil), f.122 *M. pelagica* (Atlantic Ocean – several dates; New Zealand, 23 April 1770), f.124 *M. circinnata* (Sting Rays bay, New Holland), f.126 *Phyllodoce velella* (Atlantic Ocean, 7 October 1768; southern ocean, several dates), f.134 *Beroe marsupium* (Atlantic Ocean), f.135 *B. bilabiata* (Atlantic Ocean), f.137 *B. incrassata* (Atlantic near Tierra del Fuego), f.139 *B. carolata* (Atlantic near Brasil), f.140 *B. coarctata* (southern ocean, 2 & 6 October 1769), f.142 *B. biloba* (southern ocean, 13 April 1770).

– volume 2, f.77 *Lepas anserifera* (southern ocean, 23 October 1769), f.79 *L. anatifera* (Atlantic Ocean), f.82 *L. fascigularis* (Bay of Biscay, 7 July 1771), f.86 *L. pellucens* (surface off Brasil), f.88 *L. vittata* (on *Endeavour* between Canaries and Brasil), f.91 *L. asperata* (southern ocean, 1 October 1769, 11 January 1770), f.155 *Solen radiatus* (Java), f.158 *S. albatus* (Java).

– volume 3, f.46 *Tellina radiata* (Brasil), f.94 *T. rugosa* (Pacific Ocean near Tahiti).

– volume 4, f.28 *Donax cuneata* (Nova Cambria)*, f.73 *Venus plebeja* (Nova Cambria)*, f.81 *V. maculata* (Nova Cambria* & Brasil), f.107 *V. fimbriata* (Pacific Ocean)†, f.110 *V. reticulata* (Nova Cambria)*, f.111 *V. rigida* (Brasil), f.120 *V. dilata* (Brasil)†, f.133 *V. pectinata* (Pacific Ocean)†, f.172 *V. juvenea* (Nova Cambria)*, f.186 *V. erosa* (Pacific near Nova Cambria)*, f.201 *V. opaea* (Nova Cambria, Nova Zelandia)*.

– volume 5, f.7 v. *Spondylus gaderopus* (Pacific), f.29 *Chama calyculata* (Pacific Ocean)†, f.63 *Arca barbata* (Pacific Ocean)†, f.67 *A. modiolus* (Nova Cambria)*, f.91 *A. abbreviata* (Nova Cambria)*†, f.106 *A. duplicata* (Nova Cambria)*, f.122 *A. plebeja* (Nova Cambria)*, f.123 *A. puella* (Nova Cambria)*, f.126 *A. striatula* (Nova Cambria)*, f.130 *A. turgens* (Nova Cambria)*.

– volume 6, f.53 *Ostrea lima* (Pacific Ocean, Nova Cambria)*, f.54 *O. malleus* (Pacific Ocean, Nova Cambria)*, f.75 *O. cimplanata* (grows on ships in the ocean).

– volume 7, f.50 *Mytilus margaritiferus* (Pacific Ocean, Tahiti), f.59 *M. senilis* (New Zealand, Pacific Ocean)†, f.61 *M. jubatus* (Pacific Ocean, Tahiti)†, f.83 *M. discurs* (New Zealand), f.96 *M. durus* (Pacific Ocean, Nova Cambria)†, f.129 *Pinna dentata* (Pacific Ocean, Tahiti)†.

– volume 8, f.5 *Conus imperialis β* (Pacific Ocean)†, f.15 *C. virgo* (Pacific Ocean, Tahiti), f.45 *C. ebreus α* (Pacific Ocean, Tahiti), f.57 *C. striatus* (Pacific Ocean, Tahiti)†, f.76 *C. arenatus β* (Pacific Ocean, Tahiti), f.81 *C. asper* (Pacific Ocean)†, f.167 *C. olivaceous* (Pacific Ocean)†, f.181 *C. pulicanus* (Pacific Ocean, Tahiti).

– volume 9, f.21 *Cypraea caputserpentis* g (Tahiti), f.64 *C. achatina* (Tahiti), f.68 *C. aurora* (Tahiti), f.98 *C. pressa* (Tahiti), f.122 *Bulla imperialis* (Pacific Ocean, near Tahiti).

– volume 10, f.30 *Voluta oliva* q (Tahiti).

– volume 11, f.16 *Voluta aspera* (Pacific Ocean near Tahiti), f.26 *V. carbonaria* (Pacific Ocean, New Holland), f.88 *V. insularis* (Pacific Ocean near Tahiti).

– volume 12, none.

– volume 13, f.63 *Buccinum validum* (sea near New Zealand), f.115 *Murex tritonis* (Madeira).

– volume 14, f.3 *Turbo fluitans* (Pacific Ocean, 21 March 1769), f.17 *Helix violacea* (Atlantic Ocean between the tropics), f.19 *H. janthina* (Fig.Pict.) f.61 *Alcyonium frustrum* (Atlantic off southern America), f.62 *A. anguillare* (Atlantic Ocean near Tierra del Fuego).

LEPIDOPTERA – 1 and 2 none.

NEUROPTERA & HYMENOPTERA – f.198 *Vespa tepida* (Labyrinth Bay, New Holland), f.200 *V. spiricornis* (Stingrays Bay, N.H.), f.206 *V. humilis* (Stingrays Bay, N.H.), f.208 *V. rudis* (Labyrinth Bay, N.H.), f.215 *Apis concinna* (Stingrays Bay, N.H.), f.228 *A. astuans* (Labyrinth Bay, N.H.), f.233 *Formica medullaris* (Bustard Bay, Labyrinth Bay, N.H.), f.234 *Formica viridis* (Bustard Bay, Labyrinth Bay, N.H.).

DIPTERA & APTERA – f.72 *Podura maritima* (Bay of Biscay), f.93 *Pediculus procellaria* (on *Procellaria crepidata*, Atlantic Ocean), f.94 *P. diomedea* (on *Diomedea*, S. Atlantic Ocean), f.95 *P. clypeatus* (on *Phaetontis* & *Procellaria*, Pacific Ocean), f.100 *Acarus motacillae* (on *Motacilla avida*, 70 nautical miles off Cape Blanco, Africa), f.101 *Acarus phaetontis* (on *Phaetontis* in southern Ocean), f.113 *Cancer quadratus* (Funchal, Madeira), f.116 *C. ocellatus* (New Holland), f.118 *C. pelagicus* (New Holland), f.121 *C. bulla* (Bustard Bay, New Holland), f.123 *C. natatilis* (New Holland), f.125 *C. depurator* (Atlantic Ocean); f.136 *C. caerulescens* (tropical Atlantic Ocean); f.142 *C. amplectans* (Atlantic Ocean off Brasil), f.145 *C. fulgens* (off Brasil at surface), f.150 *C. crassicornis* (off Brasil at surface), f.153 *C. gregarius* (off Patagonia), f.165 *C. vitreus* (off Brasil at surface), f.182 *Monoculus piscinus* (on *Scomber* Pelamid Atlantic Ocean), f.192 *Carcinium opalinum* (near France, Atlantic Ocean), f.194 *C. macrouram* (near France, Atlantic Ocean), f.198 *Onidium gibbosum* (near Portugal, Atlantic Ocean, inside *Dagysas*), f.202 *O. oblongatum* (Atlantic Ocean, inside *Dagysas*), f.206 *O. spinosum* (Atlantic Ocean), f.213 *Oniscus chelipes* (in algae off France, Atlantic Ocean).

HEMIPTERA – f.2 *Blatta domestica* (in Madeira culinis – not described).

COLEOPTERA – vol. 3 f.139 *Meloe ruficollis* (in the ship August 26, 1768 – the day the *Endeavour* sailed!).

The Mollusca volumes in particular are difficult to interpret with regard to *Endeavour* material. Many slips deriving from the Portland collection and also Banks's collection are localized simply Pacific Ocean and in general these are not itemized above. Solander's work on the Portland shells was undertaken late in his life and two entries with 'Habitat in Oceano pacifico' are also dated 1780 (Mollusca vol. 11, f.48 *Voluta decorata*, and f.63 *V. fuliginosa*). Clearly these specimens could have been collected on the *Resolution*, or some other voyage to the Pacific. The point is reinforced by two entries in Mollusca volume 4 (f.157 *Venus peregrina* and f.162 *V. antiquata*) where the entry is for Pacific Ocean and 'Novam Cambriam' respectively and the source of the specimens is J.R. Forster which is deleted and JB (= Joseph Banks) substituted. These specimens were therefore collected by Forster on the *Resolution* and were sold (or given) to Banks for his collection. The locality Nova Cambria which occurs in one of these cases was a later usage than the *Endeavour* voyage when the eastern Australian coast was usually refered to as New Holland. It might therefore be inferred that all the references to Nova Cambria in the slip catalogue relate to post-*Endeavour* voyage specimens. The entries for Nova Cambria, mostly in volumes 4, 5 and 6, are asterisked in the above summary (*).

Parenthetically, it can be said that the notes on *Phyllodoce velella* . . . are dated 20 August 1772, with position 59°44'N, 10°10'W of London. They thus derive from three large specimens collected by Banks on that day from his small boat which he launched during a calm, about 90 miles southwest of the Faroes. This was much further north than he expected to find the species which in his experience came no higher than Mediterranean latitudes. [*See* Banks, Joseph 1772, Journal of a Voyage to the Hebrides, Iceland and the Orkneys, BM(NH) General Library, typescript copy by H.B. Carter, f. 39.]

The possession of specimens in the Banks or Portland collections does not necessarily prove that they were part of the *Endeavour* collection, even if they have the general locality associated with one of the places visited on the voyage, because both collectors certainly received later-collected material. In the case of Solander's personal collection, however, it can be fairly assumed that if he possessed a shell (and Banks also had the same species) then it was an *Endeavour* specimen. The incidences of this are denoted in the above lists by a dagger (†).

COPIES OF ORIGINAL MANUSCRIPTS

Fair copy of descriptions of fishes and other animals obtained on Cook's First Voyage. 401 p. 20 cm. (D. & W. 41)

This is a bound volume of transcriptions of the first three zoological manuscripts discussed above (D. & W. 40a–c), the handwriting having been identified as that of Banks's Amanuensis B by Marshall (1978). It follows the original closely although the writer adopted a more disciplined use of underlining and has rearranged certain sections within some descriptions to produce a more consistent layout. In both P.A. and P.N.H. the entries for birds and crustaceans have been copied, but the bird descriptions in P.A.O.P. are not reproduced. None of the tipped-in sheets of vocabularies in the original are copied. The only major difference from the original sections (apart from the omission of A.J.C. and the notes on Icelandic fishes and birds) is that each separate description commences on the recto of a new leaf even if it continues for two following pages. This may merely be an aspect of the greater discipline in layout, but it could have resulted in a loose-leaf system capable of resorting into systematic or alphabetical sequences.

Copies of Solander's Descriptions of Animals, made during Captn. Cook's First Voyage [loose title-page, sheet 23 cm]. (D. & W. 42)
A fair copy of the descriptions of animals observed during Capt. Cook's first voyage 512 p. 32.5 cm.

Both titles refer to the same manuscript which is a later copy of Solander's descriptions of animals made during the *Endeavour* voyage. The recorded pagination is the total in a later hand, but at some time it has been reorganized with a double sequence of numbers between f.133 and f.279 – the numbers were probably inserted by Averil Lysaght in the 1950s for she re-ordered the sheets, and indexed the manuscript then.

This manuscript is a copy of many of Solander's notes on marine animals and land birds caught at sea. It comprises descriptions of birds p.1–123, reptiles p.125–131, the Linnaean class Amphibia Nantes (sharks, rays, trigger fishes etc) p.133–197, fishes p.199–277, and invertebrates p.279–511. It is arranged in accordance with the twelfth edition of Linnaeus's *Systema Naturae* (1766–1767), with hitherto unrecognized species marked for insertion between two numbered species in that edition. However, it does not include all the marine fishes recorded in other manuscripts notably P.A.O.P. (D. & W. 40c), although all the fishes recorded in the *Slip Catalogue* (D. & W. 45) Pisces 1 and 2 which were those caught at Madeira, in the tropical Atlantic and at Rio de Janeiro are included, and isolated species from the Pacific (*Raja aquilla* from Murderer's Bay, New Zealand, *Scomber lanceolatus* and *Scomber thynnus* from the tropical Pacific Ocean, and *Squalus glaucus* from Osnabrugh Island also tropical Pacific) are included. Of these only *Scomber thynnus* from Osnabrugh Island is included in P.A.O.P. in which manuscript it has been deleted.

For birds again most of the entries are copies of those descriptions in the *Slip Catalogue* (S.C. Aves), while none of those described in the three Pacific Ocean manuscripts (P.A., P.N.H., and P.A.O.P.) have been transcribed. However, there are entries in this manuscript which are not now represented in the *Slip Catalogue* notably the Solander petrel genera *Procellaria* and *Nectris* for which there are no sheets in S.C. This may, of course, mean no more than that they were lost, or removed from the *Slip Catalogue* at some time after the copy was made.

The entries for invertebrates contain a heavy preponderance of Atlantic Ocean specimens described but some Pacific Ocean entries are included. These can only have been copied from the *Slip Catalogue*, because there are only six (mostly crustaceans in the genus *Cancer* according to Solander) included in the other manuscripts (P.A.O.P. and P.N.H.) and significantly none of these are included in the present manuscript.

The conclusion therefore is that this copy of Solander's notes was made primarily from the loose slips which he employed for his *Slip Catalogue*, although limited use was made of the manuscripts from the Pacific part of the voyage. This copy was therefore a bringing-together of the scattered *Endeavour* notes in the *Slip Catalogue* because until the expedition reached the Pacific Ocean Solander had recorded his zoological notes on slips, not in a chronologically sequenced journal such as he used in New Zealand, Australia, and the islands of the Pacific.

This manuscript has been said to be incomplete by Marshall (1978) who wrote 'Many sheets are obviously missing'. Certainly, the manuscript was at one time disordered until in the 1950s Averil Lysaght put parts of it into order following Linnaeus's *Systema Naturae*, but there is no evidence that any part of it is missing. It is certainly incomplete in that it is a copy of only part of Solander's notes from the *Endeavour* voyage, but there is no evidence that it was ever intended to form a complete record of the zoology of the voyage.

Possibly Marshall was influenced to make this claim because other authors have claimed that this manuscript copy was 'lost' for many years. The origin of the statement was Gregory Mathews (1912–1913) who wrote, 'The MS. was Banks' property, and was mislaid until I discovered it in the British Museum', supplemented by Iredale (1913) who repeatedly asserted that the manuscript had been 'thrust into some corner' until Mathews's persistent enquiries had revealed it put away and labelled as 'Copies of Solander MSS'. Surprisingly for such a careful worker Lysaght (1959) repeated these statements that it had been mislaid and rediscovered. This manuscript was cited extensively by Kuhl (1820), by Gray in 1871 (see Iredale, 1913), by Sharpe (1906), and was listed in the *Catalogue of the Books, Manuscripts, Maps and Drawings in the British Museum (Natural History)* 5 (1915), and it seems to have been well enough known at the time Mathews claimed to have rediscovered it.

A manuscript copy of Solander's notes on fishes is in the Bibliothèque Centrale du Muséum, Paris, (MS 1109). This transcription (of which I have seen a microfilm) comprises 358 folios and is alphabetically arranged by genus to include a total of 245 species. The handwriting does not appear to be the same as any of Banks's staff or amanuenses (J.B. Marshall, pers. comm.). It is probably in the hand of a clerk employed to take a copy. It derives descriptions from all the known manuscripts – P.A., P.N.H., P.A.O.P. (D. & W. 40a, b, c) from C.S.D. (D. &. W. 42), and the *Slip Catalogue* (Pisces 1 & 2) (D. & W. 45), and is therefore the only place in which all the descriptions of fishes from both the Atlantic and the Pacific Oceans are gathered together.

A copy of Solander's notes, and copies of many Parkinson drawings were made for Cuvier (the drawings by Mrs Bowdich) and were cited by both Cuvier and Valenciennes in their *Histoire naturelle des Poissons* (1828–1848). This Paris manuscript is probably the one made for Cuvier.

The original manuscripts and the Banksian fair copies were kept in Banks's library in close proximity to the drawings. They were catalogued by H.F. Cary and H.H. Baber in January 1832 (see Diment & Wheeler, 1984) and all the zoological manuscripts can now be recognized except for a quarto *Catalogues of South Sea fishes*, which is no longer present in the collection. The manuscripts were later with Robert Brown in the British Museum and thus still in close association with the drawings. Not until the late nineteenth century were the zoological manuscripts transferred to the Zoological Department, and the major one *Original descriptions of fishes and other animals* (D. & W. 40) was accessioned in that Library in 1875.

The Collection as a Zoological Resource

At the time of the *Endeavour* voyage (1768–1771) the state of systematic zoology was represented by the twelfth edition of Linnaeus's *Systema Naturae* (1766–1767). A copy of this work was taken on the voyage and was clearly used as a 'field guide', albeit an incomplete guide and valid only for that part of the voyage in the Atlantic Ocean. Possibly it is this copy which is now preserved in the Zoology Library of the British Museum (Natural History) bound with interleaving and copious notes in Solander's handwriting, although I now incline to the view that these annotations were made after the conclusion of the voyage. Nevertheless this edition of Linnaeus was the latest state of the art when the *Endeavour* sailed. As such it can now be seen to be very imperfect, being especially poor in its representation of the tropical marine and terrestrial faunas in general and of the biota of the Indian and Pacific Oceans and their marginal landmasses in particular.

With the return of the expedition there is clear evidence that the primary objective of Joseph Banks was the production of a series of books describing the botany of the voyage (Carter *et al.* 1981, Diment *et al.* 1984). Although a great deal was accomplished towards this objective it finally foundered probably from several causes, amongst them the death in 1782 of the naturalist Daniel Solander (Wheeler, 1984*a*, *b*), the spiralling cost of book production, the political situation in Europe in the late eighteenth and early nineteenth centuries, and the increasing preoccupation of Banks with other affairs. Possibly had the projected great botanical work been finished and published Banks might have turned to the preparation of a comparable book on the animals of the *Endeavour* voyage. However, this seems never to have been seriously considered. Unlike the botanical drawings none of the animal drawings were finished or copied by Banks's later team of artists, and so far as is known, none was ever engraved on copper plates as a first step towards printing. The most that was attempted was the reproduction of Solander's zoological manuscripts by copyists (see Diment & Wheeler, 1984, numbers 41 and 42) but these copies were more likely to have been made as an 'insurance' against the loss of data in the event that the originals were destroyed. Of the two, number 42, the *Copies of Solander's Descriptions*, was the nearly complete compilation of Solander's notes made on the voyage presented in an organized and disciplined manner such as would be needed were the results to be collated for eventual publication. However, these notes were themselves incomplete for many of the animals, especially the invertebrates, which were collected during the voyage. It seems that so great was the concentration on the botanical collection and artistic and literary results that Solander never seriously worked on the animals collected. Not even the most dramatic of all, the kangaroos, even though a manuscript description was prepared and skins and at least one skull were brought to England, were ever formally described and published by Solander or Banks.

Solander's later career as Banks's librarian and scientific aide and his increasing responsibilities within the British Museum left less and less time for work on the zoological material. In addition, he appears to have suffered from a disposition which rendered him unable to deny help when called upon by others. Thus, after the *Endeavour* voyage he continued his collaboration with John Ellis (1710–1776) and in 1774 and 1775 was working on Ellis's, and John Fothergill's corals (and probably those collected during the voyage), while between January 1778 and June 1779 he was working on the shell collection of the Duchess of Portland. These distractions, as well as the voyage to Iceland with Banks from 12 July to 29 October 1772, must have all contributed to the gradual loss of impetus that the study of the *Endeavour* collections suffered.

The few tangible results from the *Endeavour* expedition may have contributed to the adverse comments made about Solander by later authors. The comments of Smith (1821) were especially harsh, but Smith was only 23 years of age when Solander died and may never have met him. Moreover, his comments (which perhaps significantly were not published until after Sir Joseph Banks was dead) were probably prompted from a desire to disparage the naturalist who had so closely worked with Banks in his most vigorous years. Later authors parroted Smith's criticisms, and Boulger (1898) added that Solander 'published nothing independently', which was quite untrue (Wheeler, 1984*b*) but served to advance the prejudice shown by this and several later authors.

However, whether as the result of a deliberate policy decision, or stemming from Banks's natural generosity in making available his collections and library to competent workers, the zoological material was studied from soon after the return of the *Endeavour*. Two such workers were outstanding. P.M.A. Broussonet (1761–1807) visited England for two years from 1780 and worked on Banks's collection and that of the British Museum especially on fishes. Broussonet's main publication from this collaboration was the first decade of the *Ichthyologia* (1782) in which ten species were described from Cook's voyages, some of which were from the *Endeavour* expedition. This was clearly intended to be the first of several decades under this title but no others were published. However, Broussonet is believed to have identified many of the fish drawings by writing binominal names on the sheet. Moreover, he communicated a list of names in the genus *Chaetodon* s.l. to J.F. Gmelin who published them in the thirteenth edition of the *Systema Naturae* (Gmelin, 1789). Several of these were manuscript names which derived from the *Endeavour* voyage. Later, Broussonet had a number of fishes from Banks's collection at the Faculty of Medicine in Montpellier which by 1828 had been sent to the Muséum National d'Histoire Naturelle in Paris where Cuvier, and later Valenciennes used them to supplement the notes made from the Solander manuscripts and the copies of the drawings while they were writing the *Histoire naturelle des Poissons*. Many of these preserved fishes are still in the Muséum in Paris (23 specimens are listed by Bauchot (1969)). Bauchot, in her valuable study of this collection, showed that these came from both the first and second Cook voyages, and also that there were several from Jamaica (which was not visited during the Cook voyages). Either Banks had received specimens from Jamaica (perhaps from his botanical collector, Roger Shakespear), or these specimens actually came from the British Museum collection, for Shakespear had been the collector of a large number of Jamaican fishes described by Solander in the British Museum.

Another scholar who used the Banksian collection, this time of arthropods, was J.C. Fabricius (1745–1808). A pupil of Linnaeus, he met Solander in London in 1767. At that time Solander was engaged in cataloguing the British Museum insect collection, and was actively collecting insects in the vicinity of London (Wheeler, 1984*a*) – work which might have been stimulated by Fabricius's presence. From 1772–1775 Fabricius spent each summer in London and in his autobiography (Hope, 1845) Fabricius recorded 'My friends Mr. Banks and Dr. Solander had returned from their voyage round the world, and had brought with them innumerable specimens of natural history and insects. I now lived very pleasantly. With Banks, Hunter and Drury, I found plenty of objects to engage my time, and every thing which could possibly be of service to me.' At Easter 1775 his *Systema Entomologiae* (Fabricius, 1775) was published, a work which in its perception of insect classification and description replaced the insect section of Linnaeus's *Systema Naturae*. Of approximately 1500 new taxa in the *Systema Entomologiae* about 500 were described from specimens in Banks's collection (Zimsen, 1964), many of which can be associated with the *Endeavour* voyage and others can be found in Fabricius's later works.

Where drawings of the animal exist it is difficult to be certain whether Fabricius worked with the drawing, or the specimen or specimens from which the drawing was made, or perhaps from both drawing and specimen. In the case of the feather mite and many of the crustaceans he described, which were figured by Parkinson, it is probable that he saw the drawing. In any event, no specimens of these taxa from the *Endeavour* voyage can be found today so the drawings are the only available evidence of the features of the animal. They thus have some status as types for taxonomic purposes.

This assumption that Fabricius made use of the artistic materials in Banks's collection is strengthened by the many names of Fabrician species which have been added to the drawings of Lepidoptera and Coleoptera in the British Museum (Prints & Drawings 199a8 and 199*BL) which Wheeler (1983) suggests were written by Fabricius in identifying the insects. Several of the species concerned were described as 'Mus. Dom. Banks' and this probably referred to the drawing by Parkinson in these collections.

Several birds were described from *Endeavour* material while the collection was still in Joseph Banks's ownership, mostly by John Latham (1740–1837). For example, *Cyanoramphus zealandicus* was described by Latham (1781) as the 'Red-Rumped Parrot' based on 'a fine specimen . . . now at Sir Joseph Banks's', and Latham (1785) described the Black-billed

Tropic Bird from a specimen in Banks's possession. However, Latham also made use of the drawings, perhaps in cases where no specimens existed and his account of the Frigate Petrel is based on the drawing in the collection (Latham, 1785; 1790).

The French artist and naturalist C.A. Lesueur (1778–1846) who, with François Peron, laid the foundations of the study of the Phyllum Cnidaria, examined the *Endeavour* drawings of medusae and on some wrote notes, referring to their classification. These notes on the drawings are quoted in the annotations in the Catalogue. Lesueur visited England in 1815 when he was on his way from France to North America at the commencement of his exploration of the natural history of that continent. Whether Lesueur saw only the drawings is not known, but he must have examined them in Banks's Library at that time, when Solander's manuscripts were also kept there (but Solander had been dead for thirty years at the time of his visit). Although a specimen of *Physalia physalis* is believed to be still in existence from the voyage it seems unlikely that many medusae had been preserved, or survived in preservatives from the voyage.

Another naturalist who studied the drawings and used them for descriptions of new animals was Heinrich Kuhl (1797–1821). Kuhl (1820) described several species of petrel (Procellaridae) using the Parkinson drawings as the basis for his descriptions (see, for example numbers 23 and 24 this catalogue). He appears to have cited only the drawings and made no reference to the Solander manuscripts, but in several cases adopted Solander's unpublished binominal name. From the date at which Kuhl's names were published it is clear that he saw the drawings while they were still Banks's property and kept at Soho Square.

Other naturalists during Banks's lifetime undoubtedly enjoyed the use of the drawings but have left little evidence of the use to which they put them. Some received information from Banks directly and reproduced copies of the drawings, for example J. Macartney (1810) in his study of luminous animals quotes Banks on the light production of a small crustacean (see number 226 in this catalogue), as well as reproducing the figure of *Cancer fulgens*.

After Banks's death in 1820 the drawings became the property of Robert Brown, but by 1827 they had passed to the British Museum. They were at first kept with the Banks botanical material presumably in the direct care of Robert Brown, and as late as August 1842 John Edward Gray (1800–1875) consulted the zoological drawings and manuscripts in the Botanical Department for his essay on the fauna of New Zealand (Gray, J.E., 1843*a*). Perhaps as early as 1830 G.T. Lay and E.T. Bennett compared Parkinson's drawings of fishes from the tropical Pacific with those collected on Beechey's voyage on the *Blossom* (1825–1828) and used at least one as the basis of their illustration, see number 131 in this catalogue. Some time after this both the drawings and the manuscripts had been consulted by Johannes Müller (1801–1858) and J. Henle (1809–1885) for their definitive treatment of the sharks and rays (Müller & Henle, 1838–1841). Although their introduction was dated November 1840 this must have been written for production with the third and final Heft of the volume, of which Heft 1 was issued in 1838. This means that their visit to London was earlier than 1838 when J.E. Gray made available the collections of the British Museum including the Banks and Solander materials and Hardwicke's collection of animal drawings. A number of the drawings from the *Endeavour* voyage were annotated by Müller or Henle at the time of their visit.

At around this period the drawings and sometimes the manuscripts were studied by a number of nataturalists. John Richardson (1787–1865) identified many of the *Endeavour* drawings of fishes for his paper on the ichthyology of New Zealand (Richardson, 1843*b*), and also the Forster drawings from the *Resolution* voyage which had even greater relevance to New Zealand zoology. Several other of Richardson's publications on fishes of the Pacific region contain references to the Parkinson drawings (e.g. Richardson, 1842*a*, 1842*b*, 1848) and there is little doubt that he had examined the collection in its entirety in 1840–1841. During the 1840s George Robert Gray (1808–1872) made use of the bird drawings and cited the Parkinson drawing of the Red-tailed Tropic Bird (see number 31 in this catalogue) in his catalogue of the birds in the collection of the British Museum (Gray, G.R., 1844).

The numerous citations of drawings of fishes by Richardson and Cuvier and Valenciennes (who worked from copies of the drawings and an edited fair copy of Solander's fish manuscripts) resulted in the fish drawings becoming taxonomically more important than those of other zoological groups. This led Albert Günther (1830–1910) to examine some of them, and it is thought that a manuscript list of the fish drawings (which is stored with the volumes) was made at his instigation, although it is not in his handwriting. It is not certain, however, that Günther actually studied the whole collection, but rather those drawings which had significance to some taxonomic problem.

A contemporary of Günther's, T.R.R. Stebbing (1835–1926) examined the drawings of amphipod crustaceans in the collection in the early 1880s with Günther's permission. His comments published in the volume describing the *Challenger* expedition Amphipoda (Stebbing, 1888) showed his appreciation of the importance of these drawings which in some cases had been the material on which J.C. Fabricius had based several species names more than a century earlier. Possibly no one had studied these amphipod crustacean drawings between the examination by Fabricius and that by Stebbing.

The bird drawings from the *Endeavour* voyage (and other collections of drawings of birds) were all examined and listed by R. Bowdler Sharpe (1847–1909) who published his identifications in his essay of the bird collection in *The History of the Collections in the Natural History Departments of the British Museum* (Sharp, 1906). This was the single most important study of the bird drawings made, even though parts of the collection had been studied by earlier workers, notably Latham, Kuhl, and G.R. Gray as already discussed. His list formed the basis of part of Lysaght's (1959) later study of the bird paintings which had been in the library of Sir Joseph Banks.

However, within the middle decades of the nineteenth century the emphasis of use had changed. Where the earlier authors, up to Richardson and J.E. Gray, had used the drawings to establish new taxa or to add to faunistic information, later authors (e.g. Günther, Stebbing, and Sharpe) employed them as historical records of earlier naturalists. Thus, after a period of about a century from the voyage the drawings had assumed an archival value which they still retain.

The importance of the zoological drawings is enhanced by the virtual disappearance of most of the zoological specimens since the return of the *Endeavour*. Whitehead (1969, 1978) has painstakingly studied the dispersal of the zoological collections from the Cook voyages and showed that although much of the material from all three voyages came to Banks some of it was dispersed quite quickly while other material was kept by Banks to within a few years of his death. Unfortunately, as material from all the voyages was accumulated together, it is difficult to establish what part of it was *Endeavour* material at the time of its dispersal. The general details of the disperal of each group are given below, but further particulars should be sought in Whitehead's papers from which many of these details are culled.

MAMMALS There is little evidence that mammal specimens were brought back by the *Endeavour*, and very few were described and only four were drawn. It is known that at least one skin and a skull of a kangaroo killed at the Endeavour River landing site was kept;

possibly the skins and skulls of all three kangaroo specimens were retained. At least two and possibly all three kangaroos were dressed and eaten. Banks gave one skull to John Hunter at some time before 1790, and this was later in the Museum of the Royal College of Surgeons but was destroyed by bombing there in 1941. A second skull was drawn by Nathaniel Dance, presumably while in Banks's possession, but it is not known when the drawing was made nor can it be proved that it was an *Endeavour* specimen although it is assumed to be so. There is no evidence that the skins survived (although one was stuffed see p.33), and the whole specimen in spirit which Gray (1843*b*) attributed to having come from Cook's voyages is too small to be the smallest recorded on the *Endeavour* voyage (Wheeler, 1984*b*). It cannot be proved that any other whole mammal specimen was brought back to England.

BIRDS Banks is quoted as claiming to have brought back about 500 bird specimens (Whitehead, 1969, note 2) but this must have been an exaggeration from an expedition which yielded only 32 drawings of birds and 57 species described by Solander. It is true that some petrels were shot on numerous occasions (e.g. *Procellaria velox* was recorded on ten occasions (Lysaght, 1959)) but this would still not amount to 500 specimens. Moreover, a number of the larger birds were eaten after capture. Some of the birds were kept by Banks in whose house they were examined by Latham, Kuhl and other naturalists, but he presented New Zealand birds to the British Museum in January 1773 (which can only have been from the *Endeavour* voyage) and more New Zealand birds to Marmaduke Tunstall. There may have been alcohol-preserved birds in the donations of spirit material Banks made to John Hunter in 1791, and to the British Museum in the same year, but this is not certain. So far as can be established no *Endeavour* voyage birds have survived, which is not surprising in view of preservation techniques available in the eighteenth century other than alcohol preservation.

REPTILES There is no evidence that many reptile specimens were brought to England. Two turtles were drawn on the voyage and it would have been surprising if they were not eaten soon after capture. At least two sea snakes were described and other reptiles mentioned but it is not known if they were preserved. However, a very small aquatic turtle from Batavia is known to have been preserved and kept at Soho Square until 1791 (H.B. Carter pers. comm.).

FISHES Banks's claim (see Whitehead 1969, note 2) that 500 specimens were brought back from the voyage can be accepted in the case of the fishes, for there are lists of preserved specimens in Solander's manuscripts which show that 389 fishes from the Pacific (other than Australia) were preserved. More were certainly saved from Madeira, and possibly Brazil as well; two or three Australian specimens are still preserved in Paris. It is presumed that Banks retained most of the specimens in alcohol at Soho Square until 1791 when he gave about half to John Hunter and the remainder to the British Museum. Hunter's collection became in 1800 the Museum of the Royal College of Surgeons, and Banks's donation was known as the 'New Holland Division', an inapt title if fishes were included for, so far as is known, only two or three Australian specimens were preserved. Later (1809) the Trustees of the British Museum sold a collection of duplicate specimens (which may have included some Banksian specimens) to the Royal College of Surgeons. In 1845 the College donated 348 specimens, some from the New Holland Division, to the British Museum. It is not known how many fishes were included in these transactions, or whether they were *Endeavour* specimens. The only specimens known to me in the Museum collection, which starting together (before 1800) were reunited in 1845, one having gone to the College of Surgeons Museum either via John Hunter or through the Museum duplicates sale, while the other stayed in the British

Museum, are two specimens of *Bathystethus cultratus* (Schneider, 1801); but these can be shown to be *Resolution* voyage specimens (Wheeler, 1981).

The collection of fishes which Banks gave to Broussonet, presumably around 1780–1782 when Broussonet was in England, were later taken to Montpellier, where he held the Professorship in the Faculty of Medicine. Forty-six specimens were later transferred by Cuvier to Paris and the majority of these are still preserved in the Muséum National d'Histoire Naturelle. Bauchot (1969) listed the surviving 44 specimens; possibly as many as 10 could be *Endeavour* specimens.

Other fish specimens are believed by Whitehead (1969) to have passed into the collection of Sir Ashton Lever (1729–1788) which was known as the Leverian Museum. It is not known how he obtained material from the Cook voyages, it may have been from Banks or it may have been through 'unofficial' collections made on the voyages. In 1806 the Museum was sold by auction and the material widely dispersed. Two surviving lots of fishes were reported by Whitehead, one in the Naturhistorische Museum in Vienna, the other in the Cuming Museum in London. The Vienna specimen was probably purchased at the Leverian sale by Leopold von Fichtel acting on behalf of the Austrian emperor; the London one was certainly purchased by the natural history collector Richard Cuming, whose collection formed the Cuming Museum, Southwark. Although Whitehead wrote that he searched the Cuming Museum and 'discovered' a box of dried fishes the credit for their discovery belongs properly to Raymond Chaplin when he was Assistant Curator at the Cuming Museum in the 1950s for he first drew them to attention.

INVERTEBRATES There is abundant indirect evidence for the collection of a large number of invertebrates but no lists of specimens and such material as has survived was widely dispersed and in most cases has lost such collection data as might have been expected to be associated with it.

Various coelenterates were described and figured but there is no evidence that they were preserved and, with the exception of the siphonophore, *Physalia physalis*, already mentioned, which may be an *Endeavour* specimen, none are known to survive. The corals, which were neither figured nor described, were probably given to John Fothergill (1712–1780), later purchased by William Hunter and thence passed to the Hunterian collection of the University of Glasgow. If any survive they are now in the Zoology Department of the University with shells and insects; but most have lost any supporting data.

None of the salps described or figured are known to have survived.

The mollusc shell collection must have been extensive but there is little contemporary evidence which enables us to assess how large it was or what it comprised. As collectable items shells enjoyed wide dispersal as gifts or by sales. Some were given to the Duchess of Portland (1714–1785) and were listed from *Endeavour* voyage localities in the manuscript at the Linnean Society of London, which I believe is a copy of Solander's catalogue of the collection (see Diment & Wheeler, 1984, number 57), and in the *Portland Catalogue*, the sale catalogue of her collection. From this sale specimens became widely distributed to other collectors. William Hunter also obtained *Endeavour* voyage shells when he purchased Fothergill's collection in 1781, and these also went to the University of Glasgow.

Marmaduke Tunstall (1743–1790) also received shells from the *Endeavour* voyage, presumably from Banks, at a sufficiently early date to be able to send about two hundred shells to Linnaeus (Whitehead, 1978). Tunstall also received three bird specimens from New Zealand from Banks. The shells may be presumed to be in the collection of the Linnean

Society of London, which owns Linnaeus's personal collections, but they are not identifiable as *Endeavour* specimens.

Banks's personal collection of shells was presented by him around 1815 to the Linnean Society of London where it formed part of the general museum collection maintained by the Society. By 1863 the Society's policy regarding collections changed and it was decided to retain only the Linnaean collections and some other plant collections of especial value. The Banks shell collection was given to the British Museum in 1863. Thereafter the cabinet containing the collection lost its identity and was wrongly attributed to another collection until in 1953–1954 the late Guy Wilkins was able to establish its identity as Banks's shell cabinet. Some of the shells are certainly *Endeavour* specimens, others may be, while many are clearly of later acquisition. Full details were given by Wilkins (1955).

The arthropod collections were also distributed by Banks to William Hunter and there is a considerable collection of insects in the Hunterian Collection of the University of Glasgow. Banks's remaining personal collection was given to the Linnean Society between 1811 and 1815, and from thence went to the British Museum. It is today maintained as a separate collection within the Department of Entomology, comprising 11 drawers of Coleoptera and 39 drawers of other orders. The Crustacea have been separated (they are all dry pinned specimens) and are maintained in the Department of Zoology of the British Museum (Natural History).

The insect collections contain many of Fabricius's type specimens of species described from Banks's collections (no crustacean types have been located). Both the insect and crustacean collections now suffer from imperfect labelling in the past and it is in some cases only possible to relate the specimen to locality and collection with the prior knowledge that a specimen of that taxon was captured and should be represented in the collection.

The *Endeavour* animal collections as a whole have been greatly diminished in value by the loss of labels and indentifying data, and in numbers by wide dispersal and in many cases neglect during the nineteenth century. Three significant sets of the original collections remain, the insects, the shells, and the fishes. Both the insect and shell collection deteriorated in value as a result of being inadequately labelled and then mixed with later material received by Banks, and then being added to the poorly-curated Linnean Society collection. By the time they reached the British Museum in 1863 the worst of the damage was done and the importance of the specimens individually could only be established by patient investigation. The shell collection was even completely misallocated within a century of its receipt. The fish collection now survives only as a small fraction of the original 400 or so specimens and it is dispersed though the collection of spirit-preserved material. Perhaps 25 specimens, can be located with certainty in the British Museum (Natural History) and possibly 10 others in Paris.

The history of these collections, which contrasts strikingly with the large amount of botanical material still preserved, heightens the value of the artistic and manuscript record of the voyage. In many cases, for animals described from *Endeavour* material the only extant evidence for the identity of the taxon may be found in the drawings which simply as a record of the animal's appearance show the critical features necessary for identification. The drawings offer a visual record which Solander's handwritten Latin descriptions fail to communicate, even when Solander had by chance described the precise details needed for identification. The drawings by Sydney Parkinson, Alexander Buchan, and Herman Spöring are thus not only an important visual record of the animals seen on one of the world's great voyages of discovery, but also an important zoological resource the use of which is a necessity for the correct interpretation of many early concepts in eighteenth and nineteenth century zoology.

THE CATALOGUE

This catalogue is essentially a list of the zoological drawings made on the *Endeavour* voyage. Although reference is made to Solander manuscripts and the Dryander manuscript *Catalogue* of the Banksian zoological drawings, and to the survival of cognate specimens, they are only discussed with reference to the appropriate drawing. The entries in the catalogue follow a standard format.

Explanation of a Catalogue Entry

SAMPLE ENTRY

1.(**1**:1) *Nycticebus coucang* (Boddaert, 1785) Lorisidae

DRAWING: pencil outline on branch; *r.* [pencil] 'S. Parkinson'; *v.* [pencil] 'Lemur tardigradus/[ink] Princes Island'. 369 × 270.

MANUSCRIPT: Solander – none. Dryander – Catalogue f.5 as L[emur] tardigradus dormiens —— Batavia, S. Parkinson.

NOTES: Dryander's 'L[emur] tardigradus dormiens' can be presumed to refer to the present drawing. There are no notes by Solander in his manuscripts referring to this animal; probably he never saw it, but he did briefly describe (D. & W. 45; S.C. Mammalia f. 14) *Lemur murinus* and noted that a picture existed; this is presumed to refer to the Stubbs drawing of a lesser mouse lemur, *Microcebus murinus* (J. F. Miller, 1777) which was kept alive in London by Marmaduke Tunstall (Egerton, 1976; Rolfe, 1984).

Lemur tardigradus Linnaeus, 1758, which name was used for this drawing, is confined to India and Sri Lanka.

THE HEADING: '1.(**1**:1) *Nycticebus coucang* (Boddaert, 1785) Lorisidae' represents a serial number, running through the whole collection (recently marked on the drawings by a circled number), the present volume and folio numbers, the modern scientific name of the animal represented so far as identification of it has been possible, and the family name.

'DRAWING: pencil outline on branch; *r.* [pencil] "S. Parkinson"; *v* [pencil] "Lemur tardigradus/[ink] Princes Island". 369 × 270.' represents a short description of the drawing and the medium used, and an exact transcription of annotations on the drawing with notes on the medium in which these annotations are written. *r.* = recto; *v.* = verso; line endings are indicated by an oblique stroke. The size of the paper for the drawing in mm, width × height (the animal viewed in a natural posture). Scientific names in the annotation are not italicized.

'MANUSCRIPT: Solander – none. Dryander – Catalogue f.5 . . .,' contains citations to the animal represented in the various Solander manuscripts which are cited by the numbers in Diment & Wheeler's (1984) catalogue of Solander manuscripts e.g. (D. & W. 45). These manuscripts are also cited by abbreviations of their titles as follows:

(D. & W. 40a) P.A.: *Pisces Australiae*
(D. & W. 40b) P.N.H.: *Pisces etc Novae Hollandiae*
(D. & W. 40c) P.A.O.P.: *Pisces & Anim. caetera Oceano Pacifici*
(D. & W. 40d) A.J.C.: *Animalia Javaniensia & Capensia*
(D. & W. 41) F.C.: *Fair Copy of Descriptions of Fishes.*
(D. & W. 42) C.S.D.: *Copies of Solander's Descriptions of Animals . . .*
(D. & W. 45) S.C.: *Solander's Slip Catalogue* (entries are followed by the name of the animal group in the volume, and volume number, where needed).

Dryander's manuscript *Catalogue of the animal drawings in the Banks collection* is also quoted, with folio number, the name of the animal, the locality ascribed to it and the artist's name. Again, manuscript scientific names are not italicized but if they were already published names they are printed in italic.

NOTES: under this heading may be given discussion of the identification of the specimen, references to citations of the drawing or manuscript relating to the drawing, and the possible continued existence of specimens.

The catalogue is arranged in the order in which the drawings were bound (probably by Dryander in the late 1700s) and follows the systematic sequence of the *Systema Naturae* (1766–1767). A classified index using a modern systematic sequence is given following the catalogue. An alphabetic index of current and manuscript names follows the classified index.

THE CATALOGUE

1.(**1**:1) *Nycticebus coucang* (Boddaert, 1785) Lorisidae

DRAWING: pencil outline on branch; *r.* [pencil] 'S. Parkinson'; *v.* [pencil] 'Lemur tardigradus/ [ink] Princes Island'. 369 × 270.

MANUSCRIPT: Solander – none. Dryander – Catalogue f.5 as L[emur] tardigradus dormiens —— Batavia, S. Parkinson.

NOTES: Dryander's 'L[emur] tardigradus dormiens' can be presumed to refer to the present drawing. There are no notes by Solander in his manuscripts referring to this animal, probably he never saw it, but he did briefly describe (D. & W. 45; S.C. Mammalia f. 14) *Lemur murinus* and noted that a picture existed; this is presumed to refer to the Stubbs drawing of a lesser mouse lemur, *Microcebus murinus* (J. F. Miller, 1777) which was kept alive in London by Marmaduke Tunstall (Egerton, 1976; Rolfe, 1984).

Lemur tardigradus Linnaeus, 1758, which name was used for this drawing, is confined to India and Sri Lanka.

2.(**1**:2) *Dasyurus hallucatus* Gould, 1842 Dasyuridae

DRAWING: pencil outlines with colour notes; *r.* [pencil] 'Viverra'; *v.* [pencil] 'The upper part of the body brown ash colour mix^{t} w^{t} black especially among the spots which are white, underpart of the body/pale ash colour the underpart of the tail & furthermost half of the upper part dark brown almost blk eyes black/nose & eyelids fusca'. 356 × 525.

MANUSCRIPT: Solander – none. Dryander – Catalogue f.13 as Viverra —— N.C. (= Nova Cambria), S. Parkinson, unfinished drawing.

NOTES: in a most perspicacious published study, Mahoney & Ride (1984) have shown that this drawing must have been made from a specimen captured in what is now Queensland (probably the Endeavour River landing). The drawing, although poor, with the colour description is referable to *Dasyurus hallucatus* Gould, 1842.

The drawing is of an animal which from the description in Hawkesworth (1773) formed the basis of the name *Mustela quoll* Zimmermann, 1783, a name which hitherto has been regarded as synonymous with *Dasyurus viverinus* (Shaw, 1800). Mahoney & Ride (1984) in the interest of zoological nomenclatural stability have proposed the suppression of Zimmermann's name *Mustela quoll.* Quoll was the vernacular name recorded by Hawkesworth (1773) and the journals of Banks and Parkinson use variants of this spelling which are similar to the names in present use for *Dasyurus* sp. by the Kokoimudji aboriginals whose tribal boundary extended northwards from the Endeavour river.

The skin of presumably this animal was preserved and was referred to by Pennant (1781) who noted that it had lost part of the face and that both the body and tail lengths were thirteen inches. He continued, "This was found near *Endeavour* river, on the eastern coast of New Holland, with two young ones" (followed by a reference to Hawkesworth's account of the voyage).

3.(**1**:3) *Macropus* sp. Macropodidae

DRAWING: unfinished pencil outline, animal crouched; *r.* [none]; *v.* 'Kanguru'. 358 × 525.

MANUSCRIPT: Solander (D. & W. 45) – S.C. Mammalia f.90, f.91–95 as Kanguru saliens. Dryander – Catalogue f.21 'Kanguru N.C., S. Parkinson'.

NOTES: the account of Kanguru saliens by Solander exists in two forms, written on three sides of a double foolscap folded to slip size (f.90), and on five successive slips (f.91–95). The foolscap draft (f.90) is the earlier, containing many corrections, the slips (f.91–95) are fair copies from the earlier draft, but are unique in that they give the measurements and weights of the three specimens examined. Marshall (1977) reproduced the three sides of the foolscap draft, but its sequence was disordered from the manuscript in order to take only two pages of text. Nevertheless, while the foolscap draft (f.90) may be the earlier it is a composite description which was based on at least two specimens (male and female reproductive organs are described). From Solander's manuscript (f.95) the three specimens can be distinguished as follows: a young female of eight pounds weight, a male, two to three years of age, of twenty-four pounds weight, and an adult male of eighty pounds weight. The only length given refers to a specimen 28 inches body length and 26 inches tail length. (This would appear to be in keeping with the smaller male animal.)

Fig. 2 *Macropus* sp. One of two Parkinson sketches of a 'Kanguru'. (Catalogue number 3.)

These weights differ from those given by other sources on the voyage, e.g. Banks's Journal for 14 July 1770 which records the weight as 38 lb. This source records the other specimens as 84 lb, obtained on 27 July 1770 and a third of $8\frac{1}{2}$ lb on 29 July. Parkinson's drawing (see no.4) of a leaping kangaroo suggests he drew the adult male of 84 lb.

One of the smaller specimens was skinned and stuffed, but possibly not stuffed until the *Endeavour* returned to England. Oliver Goldsmith (1791) described the "Kangaroo of New Holland . . . stuffed and brought home by Mr Banks was not much above the size of a hare . . ." (the first edition of this work appeared in 1774 – but I have been unable to consult a copy), while Pennant (1781) reports that the "Length of the largest skin . . . was three feet three inches from the nose to the tail: of the tail two feet nine". This skin (considerably larger than a hare) might have been a specimen from a later voyage.

The identity of the kangaroos from the Endeavour River has been lengthily discussed by many authors, of whom Morrison-Scott & Sawyer (1950) offered the most information based on Solander's manuscripts and Parkinson's drawings. Lysaght (1957), with considerable perspicacity recognized the George Stubbs portrait, which had dropped out of sight between 1773 and 1957, as the original of the Hawkesworth (1773) illustration in his official account of Cook's first voyage. More recently Carr (1983) has summarized the discussion, the result of which appears to be that the kangaroos (evidently at least two species were involved) from the Endeavour River cannot be positively identified, but Carr confuses the dates of capture of the largest and smallest specimens and this led him into further false assumptions.

4.(**1**:4) *Macropus* sp. Macropodidae

DRAWING: unfinished pencil outline, animal springing; *r.* [none]; *v.* [pencil] 'the whole body pale ash colour the ears excepting the base fine specled gray iris of the eye Chesnut./Kanguru/[ink] C. Endeavour's River [pencil] Endeavours River'. 525 × 358.

MANUSCRIPT: see no.3.

NOTES: see discussion under no.3.

5.(**1**:5) *Macropus robustus* Gould, 1841 Macropodidae

DRAWING: water-colour of skull and lower jaws by Nathaniel Dance; *r.* [ink] 'N. Dance'; *v.* [none]. 480 × 300.

MANUSCRIPT: Solander – [none]. Dryander – Catalogue f.21 'Cranium. Nat[l] Dance'.

NOTES: Morrison-Scott & Sawyer (1950) identified this skull as *Macropus robustus* and considered it to belong to the 80 lb (or 84 lb – depending on source followed) specimen killed on 27 July 1770. They claim that this skull was not the specimen which was given by Banks to John Hunter, which was later in the Museum of the Royal College of Surgeons in London, where it was destroyed by bombing one night in May 1941. These authors consider that the Royal College of Surgeons skull was probably that of the 38 lb animal shot on 14 July 1770. This presumably was Solander's 24 lb specimen although Cook recorded its weight clear of entrails as 28 lb (Sharman, 1970). (See no.3 for discussion of the Endeavour River kangaroos.)

6.(**1**:6) *Muntiacus muntjak* (Zimmermann, 1780) Cervidae

DRAWING: pencil outline of head, lateral and front views; *r.* [pencil] 'Cervus plicatus [cropped in binding] [ink] Parkinson'; *v.* [pencil] 'Cervus plicatus'. 530 × 355.

MANUSCRIPT: Solander – (D. & W. 45) S.C. Mammalia f.101–2 notes *Cervus axis* Briss ? Hab. in Java, written in a very unsteady hand, and perhaps referable to the specimen illustrated. Dryander – Catalogue f.23 'C. plicati Mss caput —— Pr. Isl., S. Parkinson'.

NOTES: this drawing seems not to have been supported by Solander notes unless the reference quoted above to the Java animal are relevant.

7.(**1**:7) *Milvago chimango* (Vieillot, 1816) Falconidae

DRAWING: finished pencil outline; *r.* [none]; *v.* [pencil] 'The colour of the beak pale blueish grey the feet a dirty grey blue./[ink] Terra del Fuego/[pencil] N°. 12 Falco'. 295 × 458.

MANUSCRIPT: Solander – none. Dryander – Catalogue f.33 lists a drawing as Falco ——— Tierra del Fuego, S. Parkinson, presumably referring to this drawing. This use of the genus name only confirms that it was not described in manuscript.

NOTES: listed by Lysaght (1959:272), and Sharpe (1906:173).

8.(**1**:8) *Cyanoramphus zealandicus* (Latham, 1790) Psittacidae

DRAWING: unfinished water-colour; *r.* [ink] 'S. Parkinson/[pencil] Psittacus varietas? V. S.N. XIII 329. n.88/[pencil] Aã'; *v.* 'N°.40 Green Peroquet/[ink] Otahite'. 364 × 265.

MANUSCRIPT: Solander – none. Dryander – Catalogue f.41 lists a drawing as Psittacus —— Otaheitee, S. Parkinson, presumably in reference to this drawing.

NOTES: listed by Lysaght (1959:272) and Sharpe (1906:173). This species was named by Latham (1790) based on the descriptions in his earlier account (Latham, 1781:249) of the Red-Rumped Parrot (to which Gmelin (1789) also referred). Latham's (1781) description was based on a 'fine specimen . . . now at Sir Joseph Banks's' although the locality was erroneously given as New Zealand. Latham's earlier account was cited by Gmelin (1789) as *Psittacus novae-seelandiae* (no.83) but the annotation on this drawing in an unknown hand refers to a variety of Gmelin's *Psittacus pacificus* (no.88).

9.(**1**:9) *Vini peruviana* (P.L.S. Müller, 1766) Psittacidae

DRAWING: finished pencil; *r.* [pencil] 'Latham p.255–59/Psittacus taitianius. S.N. XIII :329. n.91/Psitacus [unreadable as trimmed off] Forster Avinna/[ink] S. Parkinson'; *v.* [pencil] 'The face throat & breast white the rump & rect. dirty grey turng. blue towards the edge the feet &/beak a bright Orange Claws black. all the rest of the body w[t] dark Ultra. shaded w[t] P.B./like shining blue steel./[partly obscured, ? Otahite] No 3 Blue Perroquet'. 364 × 264.

MANUSCRIPT: Solander – none. Dryander – Catalogue f.39 lists a Parkinson pencil outline of Psittacus —— from Otaheitee.

NOTES: listed by Lysaght (1959:273) and Sharpe (1906:173–4). Latham (1781) in his description of the Otaheitan Blue Parrakeet refers to the notes on the behaviour of this Tahitian bird and the means by which it can be captured, but appears to have derived his morphological details from a specimen in the Leverian Museum and the description and figure of l'arimanon from Commerson's voyage published by Buffon (1779).

10.(**1**:10) *Calyptorhynchus magnificus magnificus* (Shaw, 1790) Psittacidae

DRAWING: unfinished pencil; *r.* [pencil] 'Lath. p.260 n.66'; *v.* [pencil] 'The

Fig. 3 *Cyanoramphus zealandicus* (Latham, 1790). Parkinson's drawing of a Tahitian bird. The Society Islands population was extinct by 1844. (Catalogue number 8.)

whole bird black spots on the head and on the shoulders dirty white the/breast feathers wav'd w[t] pale brown, the outer feathers of the tail scarlet & yellow/w[t] narrow facia of black. The iris dark brown the pupil black, the beak dirty white with the point of the upper mandible dark grey./Black Cocatoa'. 525 × 358.

MANUSCRIPT: Solander – none. Dryander – Catalogue f.39 lists a sketch without colours as 'Psittacus – Latham p.260 n.66 —— N.C., S. Parkinson'.

NOTES: listed by Lysaght (1959:273) and Sharpe (1906:174). Latham (1781:260, no 66) referred to the description by Parkinson (1773) of 'black Cockatoos of a large size' met with on the coast of New Holland, and also cited Hawkesworth (1773) but made no reference to this drawing or Banksian specimens.

11.(**1**:11) *Anas flavirostris flavirostris* Vieillot, 1816 Anatidae

DRAWING: finished pencil; *r.* [ink] 'S. Parkinson'; *v.* [pencil] 'The beak very dark brown ([inserted] changing gradually into) yellowish toward the base of the upper mandible the feet purple brown./the length of the Wing in the natural size 7½ inches./17. Anas antarctica./[ink] Terra del Fuego'. 370 × 270.

MANUSCRIPT: Solander – (D. & W. 45) S.C. Aves f.126–127; (D. & W. 42) C.S.D. f.1. Dryander – Catalogue f.57, as sketch without colours, Anas —— T.d.F., S. Parkinson.

NOTES: listed by Lysaght (1959:273) and Sharpe (1906:174).

12.(**1**:12) *Oceanites oceanicus oceanicus* (Kuhl, 1820) Hydrobatidae

DRAWING: finished pencil; *r.* [ink] 'S. Parkinson'; *v.* [pencil] 'The head, neck, breast & back soot colour, which Gradually/grows paler on the coverts of the Wings to their edges – /which are border'd w^t white, the large wing feathers &/the tail of the same sooty colour but shaded with M. blk/the upper coverts of the tail/& the sides pure white. the beak blk as are the Feet w^t a spot of yellow on each web./[ink] Decr. 22 1768/[pencil] P. oceanica'. 260 × 370.

MANUSCRIPT: Solander – (D. & W. 42) C.S.D. f.55 records three specimens obtained, the second of which is localized as America australi S.37° (Dec. 23, 1768) and was the one drawn. Dryander – Catalogue f.61, as sketch without colours P. oceanica mss —— Oc., S. Parkinson.

NOTES: listed by Lysaght (1959:273–4) and Sharpe (1906:174). Kuhl (1820) referred to this drawing in his original description of the species but also had examined a specimen which was then in Temminck's collection.

13.(**1**:13) *Pelagodroma marina marina* (Latham, 1790) Hydrobatidae

DRAWING: unfinished water colour; *r.* [ink] 'S. Parkinson'; *v.* [pencil] 'The throat, breast & belly white the Remiges, Retrices and beak/black the feet black on the webs marks of yellow as mark^d/out in the figure. [ink] Decr. 23. 1768/Lat. 37 South./[pencil] N° 6 Procellaria aequorea'. 265 × 368.

MANUSCRIPT: Solander – (D. & W. 42) C.S.D. f.57 gives the same locality and date. Dryander – Catalogue f.61, as sketch with colours P. a'quorea mss —— Oc., S. Parkinson.

NOTES: listed by Lysaght (1959:274) and Sharpe (1906:175). Latham (1785:410) as Frigate Petrel and (1790:826) as *Procellaria marina* referred to this drawing in Banks's collection and based his name on it.

14.(**1**:14) *Fregetta grallaria* (Vieillot, 1817) Hydrobatidae

DRAWING: unfinished water-colour; *r.* [ink] 'S. Parkinson'; *v.* [pencil] 'The large feathers of the wing, the tail, Beak & feet/are black the belly & coverts of the tail white./[ink] Decr. 23d. 1768./Lat.37. South/[pencil] No 7. Procellaria fregata'. 260 × 366.

MANUSCRIPT: Solander – (D. & W. 42) C.S.D. f.51 gives as locality Oceano America australis. Lat 37°S, and date Dec. 22, 1768. Dryander – Catalogue f.61, as sketch with colours P. Fregata L. —— Oc., S. Parkinson.

NOTES: listed by Lysaght (1959:274) and Sharpe (1906:175). Lysaght comments that part of Solander's description refers to *Fregata tropica* Gould, but this is at variance with Solander apparently describing only one specimen.

15.(**1**:15) ? *Pachyptila belcheri* (Mathews, 1912) Procellariidae

DRAWING: unfinished pencil; *r.* [ink] 'S. Parkinson'; *v.* [pencil] 'The beak a pale blueish lead colour – the legs & toes pale blue wt a cast of purple the webs dirty white./14 Procellaria turtur [ink] Feb. 1st. 1769. Lat. 59.00'. 270 × 366.

MANUSCRIPT: Solander – (D. & W. 42) C.S.D. f.65 Oceano Americas antarctico Terre del Fuego. Lat. S.59°. Dryander – Catalogue f.61 as sketch without colours P. Turtur MSS —— Oc., S. Parkinson.

NOTES: listed by Lysaght (1959:274) and Sharpe (1906:175). *Procellaria turtur* was published by Kuhl (1820) based on the illustration by Parkinson, the annotation of which is quoted, and also to drawing no.16. Kuhl had, however, also seen specimens in the Muséum in Paris, Bullock's museum, and in Temminck's collection. Lesson (1831) also cited *P. turtur* Banks pl.15 probably deriving this from Kuhl.

16.(**1**:16) *Pterodroma longirostris* (Stejneger, 1893) Procellariidae

DRAWING: unfinished pencil; *r.* [ink] 'S. Parkinson'; *v.* [pencil] 'The beak black the legs & toes pale violet, grey on the outermost toe the webs dirty white & partly grey veind wt dirty purple/22. Procellaria velox./[ink] Feb. 15. 1769. Lat. 48.27'' Long. 93'. 268 × 374.

MANUSCRIPT: Solander – (D. & W. 42) C.S.D. f.67 'Oceano australi', from ten localities and dates between 15 February 1769 and 11 April 1770. Dryander – Catalogue f.61 as sketch without colours P. velox Mss —— Oc., S. Parkinson.

NOTES: listed by Lysaght (1959:274–5) and Sharpe (1906:175). The name *Procellaria velox*, attributed to Banks pl.16 was published by Lesson (1831) as his Petrel Colombe, although this may have been derived from another source.

17.(**1**:17) *Macronectes giganteus* (Gmelin, 1789) Procellariidae

DRAWING: unfinished pencil; *r.* [ink] 'S. Parkinson [pencil] Procellaria gigantea α; *v.* [pencil] 18 Procellaria gigantea/Febry. 2nd 1769 Lat. 59.S'. 296 × 480.

MANUSCRIPT: Solander – (D. & W. 42) C.S.D. f.73 'Oceano antartico Terra del Fuego S. Lat. 58°S'. Dryander – Catalogue f.61. Sketch without colours P. gigantea Mss —— Oc., S. Parkinson.

NOTES: listed by Lysaght (1959:275) and Sharpe (1906:175). This drawing was referred to by Kuhl (1820), as was the next drawing. According to Kuhl there was a specimen in the British Museum. Latham (1785), referring to the Giant Petrel, also listed a specimen in the British Museum, and noted reports of the species 'by our voyagers at Staaten Land, Terra del Fuego and Isle of Desolation, and other places in the high southern latitudes'. The references to Staaten Land (Isla de los Estados) and the Isle of Desolation (Kerguelen) were records from the *Resolution* voyage, that from Tierra del Fuego probably referred to the *Endeavour* specimen drawn by Parkinson and described by Solander. Gmelin's (1789) name was based on several earlier literary accounts including Latham, J. G. Forster (1777), and Hawkesworth (1773). Several of these are linked with the *Endeavour* material which thus has some type standing.

18.(**1**:18) *Macronectes giganteus* (Gmelin, 1789) Procellariidae

DRAWING: unfinished water-colour; *r.* [ink] 'S. Parkinson. [pencil] Procellaria gigantea β'; *v.* [pencil] 'Mem. the feet are Gray:/[ink] Decr. 23. 1768/[pencil] 5 Procellaria gigantea'. 294 × 480.

MANUSCRIPT: Solander – (D. & W. 42) C.S.D. f.75 'Pelago Atlantico Americam australem . . . Lat 37°S' (December 22, 1768). Dryander – Catalogue f.61. Sketch with colours, a solid line under the previous entry to indicate identical nature of this drawing.

NOTES: listed by Lysaght (1959:275) and Sharpe (1906:175).

19.(**1**:19) *Procellaria aequinoctialis aequinoctialis* Linnaeus, 1758 Procellariidae

DRAWING: unfinished water-colour; *r.* [ink] 'S. Parkinson'; *v.* [pencil] '19 Procellaria fuliginosa/[ink] Feb 2nd 1769 Lat. 58'. 295 × 477.

MANUSCRIPT: Solander – (D. & W. 42) C.S.D. f.77 reported from two localities, Oceano Antarctico Terra del Fuego S Lat 58° (Feb. 2, 1769) and Oceano Aust. (Pacifico) Lat 44°35′ Long 109°2′ 23 Feb. 1769. Dryander – Catalogue f.61 as sketch without colour, P. fuliginosa Mss —— Oc., S. Parkinson.

NOTES: listed by Lysaght (1959:275) and Sharpe (1906:175). This drawing was referred to by Kuhl (1820) who also saw a specimen in the British Museum.

20.(**1**:20) *Pterodroma incerta* (Schlegel, 1863) Procellariidae

DRAWING: unfinished water-colour; *r.* [pencil] 'Procellaria/1/3 ?/[ink] S. Parkinson'; *v.* [pencil] 'Mem. the beak is black the legs & upper part of the feet/ palid white the lower part where mark'd off dark brown/the Claws black the under part of the whole bird is white./[ink] Decr. 23.1768/[pencil] N°4 Procellaria sandaleata'. 260 × 369.

MANUSCRIPT: Solander – (D. & W. 42) C.S.D. f.89 locality given as Habitat in oceano America australis. Lat.37°S (Dec. 22, 1768). Dryander – Catalogue f.61 as sketch with colours P. sandaliata Mss —— Oc., S. Parkinson.

NOTES: listed by Lysaght (1959:275) and Sharpe (1906:175). Lysaght said that it had been suggested, without giving any source, that two species were involved in Solander's description and the drawing but this is unlikely since only one date and locality are given in both drawing annotation and manuscript.

21.(**1**:21) *Pterodroma inexpectata* (Forster, 1844) Procellariidae

DRAWING: unfinished pencil; *r*. [ink] 'S. Parkinson'; *v*. [pencil] 'The Bill intirely black the iris of the eye brown pupil black./15 Procellaria [antarctica – deleted] lugens [substituted]/[ink] Feb. 1st 1769 Lat 59:00'. 300 × 478.

MANUSCRIPT: Solander – (D. & W. 42) C.S.D. f.91 reported from two localities oceano Antarctico, Terra del Fuego australi. Lat.59°S Long —W (Feb.1, 1769), and Oceano Australi. Lat.36°49′S, Long. 111°30′W (March 3, 1769). Dryander – Catalogue f.61 as sketch without colours, P. lugens Mss —— Oc., S. Parkinson.

NOTES: listed by Lysaght (1959:276) and Sharpe (1906:176). Kuhl (1820) referred to the two drawings (nos 21 and 22) in Banks's collection under *Procellaria lugens* in the synonymy of his *P. grisea* L.

22.(**1**:22) *Pterodroma inexpectata* (Forster, 1844) Procellariidae

DRAWING: unfinished pencil; *r*. [ink] 'S. Parkinson'; *v*. [pencil] 'The beak black the legs & that part of the foot next them dirty white the remainder black./15 Procellaria [antarctica – deleted] lugens [substituted]/Sketch made by mistake/ [ink] Febry. 3d. 1769 Lat.57 30′.' 295 × 476.

MANUSCRIPT: see above no.21.

NOTES: listed by Lysaght (1959:276) and Sharpe (1906:176).

23.(**1**:23) *Puffinus griseus* (Gmelin, 1789) Procellariidae

DRAWING: unfinished pencil of whole bird and details; *r*. [ink] 'S. Parkinson'; *v*. [pencil] 'The beak fuscus the lower mandible paler & blueish the feet of the same colour./23 Nectris fuliginosus/[ink] Feb.15. 1769 Lat.48:27″ Long :93″. 263 × 368.

MANUSCRIPT: Solander – (D. & W. 42) C.S.D. f.77, with reference to two captures Oceano Antarctico, Terra del Fuego S Lat 58° (Feb. 2, 1769), in Oceano aust. (Pacifico) Lat 44°35′ Long 109°2′ (23 Feb. 1769); Parkinson's drawing seems to refer to neither of these. Dryander – Catalogue f.61 as sketch without colours P. fuliginosa Mss —— Oc., S. Parkinson; it was also listed under *Nectris* on f.65.

NOTES: listed by Lysaght (1959:276) and Sharpe (1906:176). This drawing, but not Solander's manuscript, was referred to by Kuhl (1820) who published the name *Procellaria fuliginosa*. Gmelin (1789) referred to Cook's account of the Dark grey

Fig. 4 *Puffinus griseus* (Gmelin, 1789). Parkinson's drawing was referred to by H. Kuhl (1820) and was the basis of his name *Procellaria fuliginosa*. (Catalogue number 23.)

Petrel and to Latham (1785) probably deriving the former reference from Latham who described a specimen in the Leverian Museum.

24.(**1**:24) *Puffinus assimilis elegans* Giglioli & Salvadori, 1869 Procellariidae

DRAWING: unfinished pencil; *r.* [ink] 'S. Parkinson'; *v.* [pencil] 'The beak blue grey towards the back & the point black the legs & feet the same colour/as in Procellaria cyanopeda./24 Nectris munda/[ink] Feb.15. 1769 Lat.48.27′ Long. 93'. 270 × 367.

MANUSCRIPT: Solander – (D. & W. 42) C.S.D. f.115 as Oceano Australi, Lat. 48°27′ S, Long. 93° W (Feb.15, 1769), Lat. 35°8′ S, Long. 188°30′ W (Jan 6, 1770); the first specimen was clearly the one drawn. Dryander – Catalogue f.65 as sketch without colours N. munda Mss —— Oc., S. Parkinson.

NOTES: listed by Lysaght (1959:276) and Sharpe (1906:176). The drawing was referred to by Kuhl (1820) and is the sole type material of his species *Procellaria munda* (he erroneously gave the date as 25 February). The Parkinson drawing was described by Bourne (1959) who pointed out that *P. munda* Kuhl, 1820 had been declared a *nomen rejectum* by the International Commission on Zoological Nomenclature. The name is included as an unavailable name under the subspecies *Puffinus assimilis elegans* by Peters (1979).

25.(**1**:25) *Diomedea exulans* Linnaeus, 1758 — Diomedeidae

DRAWING: unfinished water colour; *r*. [ink] 'S. Parkinson'; *v*. [pencil] 'The face & throat white as mark'd of on the figure the whole body above & [below – faintly deleted] fusca pallida the belly/the feet whitish w[t] a cast of blue & the nails white./ [ink] Dec[r]. 23[d]. 1768/Lat.37. South/[pencil] N[o]. 9 Diomedea exulans'. 295 × 480.

MANUSCRIPT: Solander – (D. & W. 42) C.S.D. f.3 from Americam australem ubi Latit XXXVII circiter 100 Leucas nauticas a litore captus (Dec.23 1768) . . . (Mar 3 1769); (D. & W. 45) S.C. Aves f.151 – same dates, and for D. exulans var (f.157) 3 Feb. 1769 and (f.159) 2 Oct. 1769, 6 Jan. 1770, 11 Apr. 1770. Dryander – Catalogue f.65 as sketch with colours D. exulans L. —— Oc., S. Parkinson.

NOTES: listed by Lysaght (1959:277) and Sharpe (1906:176). Latham (1785) in his account of the size of the wandering albatross referred to a wing-span of twelve feet recorded 'in a manuscript at Sir Joseph Banks's'. He also mentioned specimens in the British and Leverian Museums, but there is no evidence that these were first voyage specimens. The wing-spans in the Slip Catalogue are 9 feet (f.152v), 10 feet 1 inch (f.158v), 10 feet 7 inches (f.159v) and 6 feet 11 inches (f.161v); the weight of the first was given as 12 pounds which Latham may have mis-read as the wing-span.

26.(**1**:26) *Phoebetria palpebrata* (Forster, 1785) — Diomedeidae

DRAWING: unfinished pencil; *r*. [ink] 'S. Parkinson'; *v*. [pencil] 'The bill intirely black, the iris of the eyes a yellow Brown the pupil black the skin that goes along the beak from the head/pale violet, clouded w[t] pale brown./13 Diomedia antarctica/ [ink] Feb. 1st 1760. Lat.59'. 296 × 475.

MANUSCRIPT: Solander – (D. & W. 42) C.S.D. f.9 from Terra del Fuego. 59°S (Feb,1,1769); (D. & W. 45) S.C. Aves f.160 with same data. Dryander – Catalogue f.65 as sketch without colours D. antarctica Mss —— Oc., S. Parkinson.

NOTES: listed by Lysaght (1959:277) and Sharpe (1906:176). Forster's (1785) description was based on a specimen obtained at 47°S. between 5 December 1772 and 13 January 1773 on the *Resolution* (Forster, 1844); a figure is in the G. Forster collection of drawings. Latham (1785) referred both to Forster's voyage (Forster, J.G., 1777) and Cook's published account (Hawkesworth, 1773) but appears to have derived most of his information about the Sooty Albatross from Forster. Latham's account was referred to by Gmelin (1789) who also quoted the Forster and Cook references, perhaps taking them from Latham, for his description of *Diomedia fuliginosa*.

27.(**1**:27) *Diomedea chrysostoma* Forster, 1785 Diomedeidae

DRAWING: unfinished pencil; *r.* [ink] 'S. Parkinson'; *v.* [pencil] 'The beak black excepting the back of the upper mandible & part of the under one which is a dirty greenish white./21 Diomedia profusa/[ink] Febry 3d 1769 Lat.57:30′'. 297 × 476.

MANUSCRIPT: Solander – (D. & W. 42) C.S.D. f.11 from Terra del Fuego, 58°31′S (Feb.3, 1769), and 48°27′S (Feb.15, 1769); (D. & W. 45) S.C. Aves f. 162 with same data. Dryander – Catalogue f.65 as sketch without colours D. profusa Mss —— Oc., S. Parkinson.

NOTES: listed by Lysaght (1959:277) and Sharpe (1906:176–7). Latham (1785) described the Yellow-nosed Albatross from a specimen taken off the Cape of Good Hope in the British Museum, but mentions the species occurring in the southern hemisphere all round the pole from 30 to 60 degrees and then specifies one 'caught in lat. 57.30.S. in the month of February'. This suggests that Latham had taken these data from the Parkinson drawing, and not from Solander's manuscripts.

28.(**1**:28) *Fregata magnificens* Mathews, 1914 Fregatidae

DRAWING: unfinished pencil; *r.* [ink] 'S. Parkinson'; *v.* [pencil] 'the Beak is of a lead colour whitish towards the base of/the upper mandible the bag is of a dirty Orange the feathers of the whole body is quite black having a/ cast of Purple on the back the feet & Claws lead Colour./To be colord from/N° in Cag N°. / Pelecanus Aquilus JB/Specimen lost N° 3/[ink] Rio Janeiro'. 295 × 480.

MANUSCRIPT: Solander – (D. & W. 42) C.S.D. f.19 from America meridionali; (D. & W. 45) S.C. Aves f.168, as Pelecanus Aquilus var., same data. Dryander – Catalogue f.67 as sketch without colour P. Aquilus L. —— Bras., S. Parkinson.

NOTES: listed by Lysaght (1959:277) and Sharpe (1906:177).

29.(**1**:29) *Phalacrocorax albiventer* (Lesson, 1831) Phalacrocoracidae

DRAWING: finished pencil; *r.* [ink] 'S. Parkinson'; *v.* [pencil] 'The beak & all the bare part around the eye is a brownish grey – the point only excepted which is whitish/the iris of the eyes grey pupil black the feet something reddish./N° 11 Pelecanus antarcticus/[ink] Terra del Fuego'. 290 × 480.

MANUSCRIPT: Solander – (D. & W. 42) C.S.D. f.15 from Terra del Fuego & adhuc australius; (D. & W. 45) S.C. Aves f.170 – same data. Dryander – Catalogue f.67 as sketch without colours, P. antarcticus Mss —— T.d.F., S. Parkinson.

NOTES: listed by Lysaght (1959:277) and Sharpe (1906:177).

30.(**1**:30) *Sula serrator* G. R. Gray, 1845 Sulidae

DRAWING: partly coloured pencil sketch; *r.* [ink] 'S. Parkinson'; *v.* [pencil] '1. Pelecanus sectator/[ink] Aehie ne Mauwe'. 295 × 475.

MANUSCRIPT: Solander – (D. & W. 42) C.S.D. f.17, habitat Oceano Australiam septentrionalem 36°–33° S, 185°–187° W, Dec.24, 1769; (D. & W. 45) S.C. Aves f.171 – same data. Dryander – Catalogue f.67 as sketch without colour, P. Bassanus L. —— N.Z., S. Parkinson.

NOTES: listed by Lysaght (1959:278) and Sharpe (1906:177) who gave the name on the annotation as *Pelecanus serrator*. The Dryander entry must refer to this drawing as it is the only *Pelecanus* sketched by Parkinson in New Zealand. Gray's (1845) original description cited the Parkinson drawing as *Pelecanus serrator*, Banks, Icon. ined. 30, and appears to have derived the name for the species from a misreading of Solander's manuscript name.

31.(**1**:31) *Phaethon rubricauda melanorhynchos* Gmelin, 1789 Phaethontidae

DRAWING: signed water-colour of bird in flight: *r.* [ink] 'Phaeton erubescent/ Sydney Parkinson pinx[t] 1769/[pencil] Tawai'; *v.* [none]. 290 × 315.

MANUSCRIPT: Solander – (D. & W. 42) C.S.D. f.29, locality given as Oceano australi ca Otaheite & Nigus, March 21, 1769; (D. & W. 45) S.C. Aves f.181–184v. same data. Vernacular name in both mss Tavai 'Otaheitensibus'. Dryander – Catalogue f.69 as finished in colour P. erubescens Mss —— Oc., S. Parkinson and sketch with colour — caput — .

NOTES: listed by Lysaght (1959:278) and Sharpe (1906:177). Latham (1785) described the Black-billed Tropic Bird from a specimen in Sir Joseph Banks's possession and gave as localities for the species 'Turtle and Palmerston Islands, in the South Seas'. Gmelin (1789) based his *Phaeton melanorhynchos* solely on Latham's account. G. R. Gray (1844) cited the Parkinson drawing and probably from it Solander's manuscript name as 'P. erubescens Banks, Icon. ined. t.31.' in his synonymy of the Red-tailed Tropic Bird.

32.(**1**:31) *Phaethon rubricauda melanorhynchos* Gmelin, 1789 Phaethontidae

DRAWING: finished water-colour of bird's head; *r.* [pencil] 'on the same Paper with the Bird'. *v.* [none]. 159 × 285.

MANUSCRIPT: see no.31.

33.(**1**:32) *Larus maculipennis* Lichtenstein, 1823 Laridae

DRAWING: unfinished pencil; *r.* [ink] 'S. Parkinson'; *v.* [pencil] 'The beak & feet the col[r] of minium – the breast & belly white w[t] a cast of red the same as in the Cocatoo w[t] the red crest/the claws dark brown – the length of the Wing in the natural size 11 inches/16 Larus gregarius/[ink] Terra del Fuego'. 265 × 364.

MANUSCRIPT: Solander – (D. & W. 42) C.S.D. f.35, locality given as Terra del Fuego; (D. & W. 45) S.C. Aves f.190, same data. Dryander – Catalogue f.69 as sketch without colour, L. gregarius Mss —— T.d.F., S. Parkinson.

NOTES: listed by Lysaght (1959:278) and Sharpe (1906:177–8); the latter quoted from Solander's manuscript notes.

34.(**1**:33) *Gygis alba candida* Gmelin, 1789 Laridae

DRAWING: pencil sketch; *r.* [ink] 'S. Parkinson/[pencil] Epérai'; *v.* [pencil] 'The whole bird intirely white the beak a lead colour, as are also the toes, the webs between white/the Rachi of the wing feathers pale brown & those of the tail black/ N°. 2. Egg Bird/[ink] Otahite'. 265 × 365.

MANUSCRIPT: Solander – (D. & W. 42) C.S.D. f.101, locality given as Otahaensibus, Oceano Pacifico . . . Insulam Otahe (July 28, 1769); (D. & W. 45) S.C. Aves f.210, same data. Dryander – Catalogue f.71 as sketch without colour, Sterna – Ot., S. Parkinson. (See notes.)

NOTES: listed by Lysaght (1959:278) and Sharpe (1906:178). This drawing has not previously been associated with any bird described in the Solander manuscripts. It is unquestionably a tern and was listed by Dryander as *Sterna* sp. from Tahiti. The only *Sterna* described from Tahiti was Sterna nigripes (see references quoted above) details of which appear to agree with the notes on coloration in the annotation. The specimen of Sterna nigripes was preserved in Cagg No 6 as bird specimen A 20.

Gmelin's (1789) *Sterna candida* was solely based on Latham's (1785) account of the White Tern which described a specimen in the Leverian Museum, although the locality given for the species was Christmas Island 'and other parts of the South Seas. Seen also off the island of St. Helena'. Latham must have derived this information from several sources although he cites no literature. Christmas Island in the Line Islands was visited on Cook's third voyage December 1777–January 1778.

35.(**1**:34) *Ptilinopus purpuratus* (Gmelin, 1789) Columbidae

DRAWING: unfinished water-colour; *r.* [ink] 'S. Parkinson/[pencil] Columba porphyraea Forster/purpurata S.N. XII:784. n.64/Oopãa'; *v.* [pencil] 'N°. 4. Green dove/[ink] Otahite'. 366 × 265.

MANUSCRIPT: Solander – none. Dryander – Catalogue f.89 – two identical entries refer to sketches with colour within *Columba* from Otaheiti and relate to this drawing and no.36.

NOTES: listed by Lysaght (1959:278) and Sharpe (1906:178). Gmelin's name was based solely on the description of the Purple-crowned Pigeon in Latham (1783). This was described from a specimen from Otaheite in the Leverian Museum. Latham records the Tahitian vernacular name as Oopa or Oopara; he also describes variation in plumage in his Purple-crowned Pigeon in specimens or descriptions from Uliatea (Raiatea) and Tonga Taboo (Tongatabu, Friendly Islands).

36.(**1**:35) *Gallicolumba erythroptera* (Gmelin, 1789) Columbidae

DRAWING: unfinished water-colour; *r.* [pencil] 'Latham 2. p.624. n.13./Columba erythroptera S.N. XIII:775. p.10./Amāhò/amehò/[ink] S. Parkinson'/[colouring directions written on the drawing]; *v.* [pencil] 'the red on the neck brighter some of a fine shiny purple/No 1 Columba pectoralis/[ink] Otahite'. 265 × 360.

MANUSCRIPT: Solander – (D. & W. 40c) P.A.O.P. Aves f.1 (261) as Columba pectoralis with reference to the drawing. Dryander – Catalogue f.89 (see entry for no.35).

NOTES: listed by Lysaght (1959:278) and Sharpe (1906:178). Gmelin's name was based solely on Latham's description of the Garnet-winged Pigeon (Latham, 1783); he recognized three varieties, α from the island of Eimeo (Moorea), β from Tahiti, and γ from Tanna, New Hebrides. These three localities all derive from Latham's (1783) account in which he reported that the Otaheite specimen was at Sir Joseph Banks's, while the Eimeo specimen (collected on Cook's third voyage) was in the Leverian Museum. The three varieties differed in details of plumage and size. This species is now extinct in the Society Islands (Lysaght, 1959).

37.(**1**:36) *Ramphocelus bresilius* (Linnaeus, 1766) Emberizidae

DRAWING: pencil sketch with colour; *r.* [ink] 'S. Parkinson/[pencil] Loxia mexicana'; *v.* [pencil] 'The whole wings & tail black a little inclining to brown the feathers of the/Back at their bases are black & their edges scarlet which makes it/ look darker than the scarlet of the Belly – is more yellow than the rest./the legs fusca the beak black excepting the oblong space mark'd of/on the base of the under mandible which is white. N°1 Preserved dry in Box N° /[ink] Rio Janeiro'. 295 × 239.

MANUSCRIPT: Solander – none. Dryander – Catalogue f.95 as sketch with colours, L. mexicana L. n.7 —— Bras., S. Parkinson.

NOTES: listed by Lysaght (1959:279) and Sharpe (1906:178). *Loxia mexicana* dates from Linnaeus (1758).

38.(**1**:36) *Turdus falcklandi magellanicus* P. P. King, 1830 Muscicapidae

DRAWING: unfinished pencil; *r.* [ink] 'S. Parkinson'; *v.* [pencil] 'N°. 11 Turdus/ [ink] Terra del Fuego'. 295 × 239.

MANUSCRIPT: Solander – none. Dryander – Catalogue f.91 listed as sketch without colour, T. —— T.d.F., S. Parkinson.

NOTES: listed by Lysaght (1959:279) and Sharpe (1906:178).

39.(**1**:37*a*) *Sporophila caerulescens* (Vieillot, 1817) Emberizidae

DRAWING: unfinished water-colour; *r.* [ink] 'S. Parkinson'; *v.* [pencil] 'N°. 2/ Cagg N° / Rio de Janeiro'. 293 × 238.

MANUSCRIPT: Solander – none. Dryander – Catalogue, not identifiable.

NOTES: listed by Lysaght (1959:279) and Sharpe (1906:178).

40.(**1**:37*b*) *Volatina jacarina* (Linnaeus, 1766) Emberizidae

DRAWING: finished water-colour, signed; *r.* [ink] 'Loxia nitens/Sydney Parkinson pinx[t] ad vivum 1768/Brasil'; *v.* [ink] 'of the Coast of Brasil/Nov[r]. 8th. 1768'. 292 × 237.

MANUSCRIPT: Solander – (D. & W. 42) C.S.D. f. 119, locality given as 'in Brasilia australi'; (D. & W. 45) S.C. Aves f. 267, same data. Dryander – Catalogue f. 95 as finished in colours, L. nitens Mss —— Bras., S. Parkinson.

NOTES: listed by Lysaght (1959:279) and Sharpe (1906:179).

41. (**1**:38*a*) *Motacilla flava* Linnaeus, 1758 Motacillidae

DRAWING: finished water-colour, signed; *r*. [ink] 'Motacilla avida/Sydney Parkinson pinxt 1768'; *v*. [ink] 'Sept^{tr}. 28. 1768/Lat. 19.00 – north'. 214 × 270.

MANUSCRIPT: Solander – (D. & W. 42) C.S.D. f. 121 as 'Habitat in Africa qua in mari Atlantico (Lat 19 N) African . . . (Sept 28, 1768)'; (D. & W. 45) S.C. Aves f. 275, same data. Dryander – Catalogue f. 103 as finished in colours, M. avida Mss —— Oc., S. Parkinson.

NOTES: listed by Lysaght (1959:279) and Sharpe (1906:179).

42. (**1**:38*b*) *Oenanthe oenanthe* (Linnaeus, 1758) Muscicapidae

DRAWING: finished water-colour, signed; *r*. [ink] 'Motacilla velificans./Sydney Parkinson pinxt ad vivum 1768 Sept./T. 10. P. 6. Sept. 4. 1768./[pencil] Oenanthe'; *v*. [ink] 'of the Coast of Spain'. 273 × 239.

MANUSCRIPT: Solander – (D. & W. 42) C.S.D. f. 123 as from Europe australiore prope litora Gallicia Hispanorum in Nave capta Sept. 3 1768; (D. & W. 45) S.C. Aves f. 277, same data. Dryander – Catalogue f. 103 as finished in colour, M. velificans Mss —— Oc., S. Parkinson.

NOTES: listed by Lysaght (1959:279) and Sharpe (1906:179).

43. (**1**:39) *Chelonia mydas* (Linnaeus, 1758) Cheloniidae

DRAWING: unfinished pencil sketch, dorsal view; *r*. [ink] 'S. Parkinson'; *v*. [pencil] 'Testudo Mydas/[ink] Endeavours River'. 266 × 371.

MANUSCRIPT: Solander – (D. & W. 42) C.S.D. f. 125 Novam Hollandium male and female, with vernacular names for both from Australia, and for the species from the islands of the Pacific; (D. & W. 45) S.C. Amphibia 1, f. 14, same data; (D. & W. 40d) A.J.C. f. 3 (303) Princes Island and vernacular. Dryander – Catalogue f. 109 as sketch without colours T. Midas L. superne —— N.C., S. Parkinson.

NOTES: clearly only the descriptions made in Australia (Nova Hollandia, or Nova Cambria in Dryander's Catalogue) refer to the Endeavour River specimens. The reports from the Pacific islands and Princes Island were either a result of communication with the natives, or of native fishing. Despite this being a well-known species described by Linnaeus mainly from carapaces, Solander gave a moderately detailed account and described them copulating in July and August.

Dr E. N. Arnold (personal communication) points out that the drawing is not sufficiently detailed for critical distinction from *Chelonia depressa*, but *C. mydas* is the more probable species.

44.(**1**:40) *Chelonia mydas* (Linnaeus, 1758) Cheloniidae

DRAWING: unfinished pencil sketch, ventral view; *r.* [ink] 'S. Parkinson'; *v.* [pencil] 'Testudo Mydas/[ink] Endeavours River'. 268 × 370.

MANUSCRIPT: see above no.43; Dryander – Catalogue f.109 as sketch without colours T. Midas L. inferne —— N.C., S. Parkinson.

NOTES: see above no.43.

45.(**1**:41) *Caretta caretta* (Linnaeus, 1758) Cheloniidae

DRAWING: unfinished pencil sketch, dorsal view; *r.* [ink] 'S. Parkinson'; *v.* [ink] 'Decr. 23 1768/Lat.37. South/[pencil] N°. 1 Testudo Caretta'. 262 × 365.

MANUSCRIPT: Solander – (D. & W. 42) C.S.D. f.127 same details as above 'supra aquam domiens, capta . . . circiter milliaria nautica a litore'; (D. & W. 45) S.C. Amphibia 1 f.16, same data, the carapace length given as 25 inches. Dryander – Catalogue f.109 as sketch without colours T. Caretta L. superne —— Oc., S. Parkinson.

NOTES: this specimen of the loggerhead turtle caught off Brazil was drawn by Parkinson from the ventral and lateral views (see below). Dryander's Catalogue lists each drawing distinguishing them as 'inferne' and 'a latere visa' respectively.

46.(**1**:42) *Caretta caretta* (Linnaeus, 1758) Cheloniidae

DRAWING: unfinished pencil sketch, ventral view; *r.* [ink] 'S. Parkinson'; *v.* [pencil] 'No.1 Testudo Caretta/Decr. 23^d. 1768. Lat.37.S./N°. 10 Testudo Caretta/ Decr. 23. 1768/Lat.37. South'. 263 × 372.

MANUSCRIPT: see above, no.45.

NOTES: see above, no.45.

47.(**1**:43) *Caretta caretta* (Linnaeus, 1758) Cheloniidae

DRAWING: unfinished pencil sketch, side view; *r.* [ink] 'S. Parkinson'; *v.* [pencil] 'N° 10/[ink] Decr. 23/Lat.37. South/[pencil] N°. 1 Testudo Caretta'. 262 × 369.

MANUSCRIPT: see above, no.45.

NOTES: see above, no.45.

48.(**1**:44) *Raja nasuta* Müller & Henle, 1841 Rajidae

DRAWING: pencil sketch; *r.* [pencil] 'Raja oxyrinchus L. ?/[ink] S. Parkinson'; *v.* [pencil] 'Clay colour with the edges of the body finns up to the nose ting'd with red./ 21. Raia nasuta/[ink] Totarra'nue'. 475 × 300.

MANUSCRIPT: Solander – (D. & W. 42) C.S.D. f.135, from Oceano Australiam . . . Totaranui; (D. & W. 45) S.C. Amphibia 1, f.153–154, same data. Dryander – Catalogue f.119 as sketch without colour R. oxyrinchus L? —— N.Z., S. Parkinson.

NOTES: Müller & Henle (1841) based their description solely on this drawing which they attribute to Banks MS 44, from Südsee, but make no reference to Solander's manuscripts. This drawing therefore has type status. Günther (1870) reproduced Solander's manuscript description in full but could not then find Parkinson's drawing (possibly because it was concealed in volume 1 with other Linnaean 'Amphibia'), although as he commented, the drawing had been seen by Richardson (1843*a*) who cited the locality, Totaeranue, volume and folio number. The drawing was reproduced by Whitehead (1968) with annotations transcribed. The length of the drawing, i.e. the pencil image, was given as 345 mm, which unfortunately was assumed by Garrick & Paul (1974) to be the length of the specimen. Although many of the *Endeavour* (and *Resolution*) fishes were drawn at life size, it would be a dangerous assumption to assume that all were, especially such potentially large animals as rays, skates, and sharks. Garrick & Paul describe *Raja nasuta*, known as the 'rough skate in New Zealand' as the most common skate on the North Island continental shelf.

49.(**1**:45) *Aptychotrema banksii* (Müller & Henle, 1841) Rhinobatidae

DRAWING: finished pencil dorsal view and detail of ventral side of head by H. D. Spöring; *r.* [pencil] 'RAJA rostrata/rostrata deleted/Rhinobatos L. ?'; *v.* [pencil] 'Rhinobates (Rhinobates) Banksii Müller und Henle'. 426 × 358.

MANUSCRIPT: Solander – (D. & W. 40) P.N.H. f.3 (85), locality not given, date 29. April 1770; (D. & W. 45), S.C. Amphibia 1, f.162, 'Habitat in Oceano Pacifico Novae Hollandia JB & DS'. (and Oceano Jamaicensis Shakespear). Dryander – Catalogue f.119 as finished without colour R. rhinobatos L. ? —— N.C., Spöring.

NOTES: Solander's manuscript (D. & W. 45) S.C. Amphibia 1, folio 162 is a post-voyage entry as it records Raja rostrata from the *Endeavour* voyage, with the reference 'Fig. Pict.' to this drawing, and then continues to include the reference to the specimen from Shakespear's collection from Jamaica with an MB indicating a specimen was in the British Museum. This presumably referred to a specimen of *Rhinobatos* from the Caribbean.

This drawing was referred to by Müller & Henle (1841) (referring to Banks MS. 45) with the locality at Neuholland; this is the sole reference for typification of the species. From the annotations on the drawing it can be seen that no locality was given, nor did Solander's manuscript account (P.N.F. f.3) although from its inclusion in this part of the manuscript (Pisces Novae Hollandiae) it clearly came from Australia (= New Holland in Banks and Solander usage, or N.C., Novae Cambria, of Dryander's *Catalogue*). The date given by Solander (29 April 1770) shows that this fish was captured in Botany Bay, New South Wales (Groves, 1962). Why Richardson (1843*a)* should have included this species in a list of New Zealand fishes is unknown, except perhaps that he may, as Garrick & Paul (1971) suggest, have misread Müller & Henle's Neuholland for New Zealand. Richardson (1843*a*) did not, however, simply copy the reference from Müller & Henle, who refer to the drawing as Banks MS 45, for he specifically cited Banks, fig. pict. 1 p.45, thus drawing attention to the drawing in the first volume, and elsewhere (Richardson,

1843*b*) he referred to the drawing as Parkinson 1, t.45. However, there is no locality on the drawing and this must have misled Richardson, although plainly he did not refer to the Solander manuscripts for these are clearly localized. Garrick & Paul (1971) have recently removed this taxon from the New Zealand faunal list.

This drawing was reproduced by Whitehead (1968).

A specimen given by Banks to Broussonet and now in the Muséum National d'Histoire Naturelle, Paris and believed to be from Australia, is not considered to be the original for this drawing on account of disparity in length (drawing length 465 mm, specimen 205 mm) (Bauchot, 1969).

50.(**1**:46) *Urolophus testaceus* (Müller & Henle, 1841) Dasyatidae

DRAWING: finished pencil dorsal view and detail of ventral side of head by H. D. Spöring; *r.* [pencil] 'RAJA testacea/N.B. the 200^{d} pounder wanted the upper fin on the extremity of/the tail, & the small fin near the stings, the head . . .'/[last line trimmed off]; *v.* [pencil] 'Trygonoptera testacea Müller und Henle'. 526 × 360.

MANUSCRIPT: Solander – (D. & W. 40) P.N.H. f.3 (85) locality not given; date 30 April 1770. Dryander – Catalogue f.119, presumably one of three finished drawings without colour of Raja —— N. C., Spöring (the others being nos. 51 and 52).

NOTES: Müller & Henle (1841) used this drawing as the sole source of *Trygonoptera testacea*, citing it as 'Banks. MS. 46'; they did not, however, apparently cite Solander's manuscript description. The drawing was again referred to as 'Parkinson, 1.t.146 (*sic*)' by Richardson (1843*b*), and again mistakenly associating the drawing with New Zealand whereas the fish was captured on 30 April 1770 in 'Sting Ray's Bay', later Botany Bay, New South Wales. This drawing was reproduced by Whitehead (1968).

According to Whitley (1940) *Urolophus testaceus* grows to only 30 inches (76 cm) in length; the note referring to another specimen of 200 pounds weight could not have referred to this species (as the note of its lacking the caudal and dorsal fins confirms). This and the other large sting rays caught in Botany Bay in May 1770 were referred to *Bathytoshia brevicaudata* (Hutton, 1875) by Whitley (1940) (= *Dasyatis brevicaudata*, of authors).

51.(**1**:47) *Trygonorhina fasciata* Müller & Henle, 1841 Rhinobatidae

DRAWING: finished pencil dorsal view, details of ventral side of head by H. D. Spöring; *r.* [pencil] 'RAJA fasciata/Long. 2 ped: 1½ unicas'; *v.* [pencil] 'Trygonorhina fasciata Müller und Henle'. 525 × 360.

MANUSCRIPT: Solander – (D. & W. 40) P.N.H. f.2 locality not given, date 29 April 1770. Dryander – Catalogue f.119, presumably one of three finished drawings without colour of Raja —— N.C., Spöring (the others being nos. 50 and 52).

Fig. 5 *Trygonorhina fasciata* Müller & Henle, 1841. Drawing by Spöring of a fish caught on 29 April 1770 at Sting Ray Bay (later Botany Bay), Australia. (Catalogue number 51.)

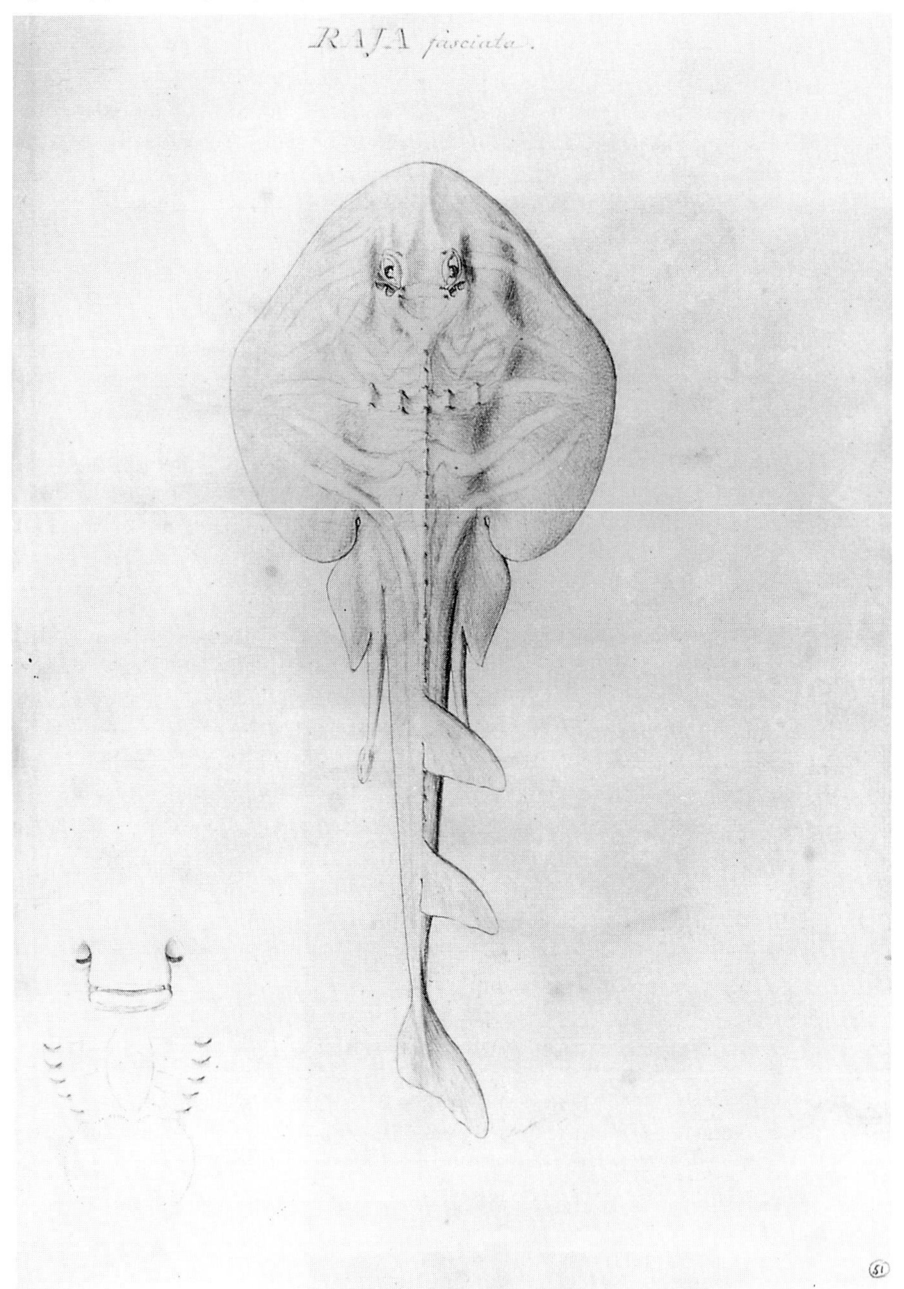

NOTES: *Trygonorhina fasciata* was based by Müller & Henle partly on this drawing, cited as Banks. MS.47, but also on a Quoy & Gaimard specimen in alcohol in the Paris Museum (cf. Garrick & Paul (1971) who claim that the species name was based solely on Banks's drawing). The locality given by Müller & Henle was

Neuholland. Despite this Richardson (1843*a*; 1843*b*) included this taxon in his list of New Zealand fishes, on the first occasion citing it as Banks, fig. pict. 1, t.47 and the second as Parkinson, i.t.47. It is possible that Richardson assumed that this fish came from New Zealand as no locality was given on the drawing, but he clearly did not refer to Solander's manuscript entitled *Pisces Novae Hollandiae* (= Australia) or the date 29 April 1770, when the *Endeavour* had just arrived at Botany Bay. Garrick & Paul (1971) have recently removed this fish from the Zealand fauna list.

This drawing was reproduced by Whitehead (1968).

52.(**1**:48) *Myliobatis australis* Macleay, 1881 — Myliobatidae

DRAWING: finished pencil dorsal view, details of ventral and lateral views of head by H. D. Spöring; *r.* [pencil] 'RAJA/macrocephala/Willugh. app. tab. 10. f.3/Long 4 pedum'; *v.* [pencil] 'Myliobates Nieuhofi Müller und Henle/ [ink] Botany Bay'. 528 × 359.

MANUSCRIPT: Solander – none; Dryander – Catalogue f. 119, presumed to be one of three finished drawings without colour of Raja —— N.C., Spöring (the others being nos. 50 and 51).

NOTES: this drawing was cited by Müller & Henle (1841) as *Myliobatis Nieuhofi*, with a reference to Banks. MS.48; their annotation appears on the verso. Richardson (1843*a*, 1843*b*) includes *M. nieuhofi* (Bloch & Schneider, 1801) in the New Zealand fauna apparently on the Müller & Henle reference but citing Raia macrocephala, Banks, fig. pict. 1, t.48 (Richardson, 1843*a*). The reference to Willughby in the annotations is to Willughby (1686) whose description, derived from Nieuhof, formed the sole basis for Bloch & Schneider's (1801) name.

53.(**1**:49) *Prionace glauca* (Linnaeus, 1758) — Carcharhinidae

DRAWING: finished water-colour, lateral view; *r.* [ink] 'Squalus glaucus./Sydney Parkinson pinx[t]: 1769./mow otaa'; *v.* [ink] 'April 10.1769. off Osnabrugh Island South Sea'. 292 × 466.

MANUSCRIPT: Solander – (D. & W. 42) C.S.D. f.187 (171) from Oceano australis . . . Osnabrugh Island, captus April 10, 1769, Lat. 17°, Length 6 ft 7 in; (D. & W. 45) S.C. Amphibia 1, ff.211–217, with same data. Dryander – Catalogue f. 121 as finished in colours S. glaucus L. e latere —— Oc., S. Parkinson.

NOTES: Solander's manuscript notes in the Slip Catalogue (D. & W. 45) exist in several states. The pencil notes on ff.216 (two leaves) are bound in reverse order, and may be presumed to be the original notes; the last page is in ink and contains detailed measurements. The measurements are repeated in ink in fair copy on f.217. A formally arranged description appears on ff.214–215 (four pages), and a short diagnosis and citation of Linnaeus's account in Solander's hand appears on f.213. The specimen was dissected, and Solander gives details of internal anatomy. The internal helminth parasite Fasciola tenacissima Solander, described on 11 April, came from this fish, see no.239.

Osnabrugh Island is now known as Mururoa, and lies south of the Tuamotu group.

54.(**1**:50) *Prionace glauca* (Linnaeus, 1758) Carcharhinidae

DRAWING: finished pencil, ink and wash dorsal view and pencil detail of underside of head; *r.* [ink] 'Squalus glaucus./Sydney Parkinson pinx^{t}. 1769'; *v.* [pencil] 'Squalus glaucus/ [ink] April 10. 1769. off Osnabrugh Island South Sea/ [pencil] The back pretty low. P.B. shaded towards the top w^{t} indian ink & D^{0} gradually/ running into the belly which is silvery lightly ting'd w^{t} red the eye clouded w^{t} black-grey the circle white ting'd wt blue'. 472 × 290.

MANUSCRIPT: Solander – see above no. 53. Dryander – Catalogue f. 121 as finished without colour, S. glaucus L. a tergo —— Oc., S. Parkinson.

NOTES: see no. 53.

55.(**1**:51) *Carcharhinus* sp. Carcharhinidae

DRAWING: finished water-colour oblique view of back of shark; details of eyes; *r.* [ink] 'Squalus carchadius./Sydney Parkinson pinx^{t} 1768/mow'; *v.* [none]. 296 × 472.

MANUSCRIPT: Solander – (D. & W. 45) S.C. Amphibia 1 ff. 206–209, no locality or date given, specimen was 5 feet 4 inches long; (D. & W. 42) C.S.D. f. 177 (159) repeats the Slip Catalogue entry. Dryander – Catalogue f. 121 as finished in colours Squalus Carcharias L. e latere —— Oc., S. Parkinson.

NOTES: Parkinson's mis-spelling of the species name must have been due to misreading a written text ('dius' for 'rias'). Solander identified this shark with *Squalus carcharias* Linnaeus, 1758, the present *Carcharodon carcharias*, but there was much confusion in the eighteenth century with carcharhinid sharks and *Carcharodon*. Solander's manuscript notes exist in two states in the Slip Catalogue (D. & W. 45); f. 209 is a large folded sheet mostly written in pencil which must be his first draft notes. They are neatly transcribed at ff. 207–208 (four pages) and f. 206 is a diagnosis with reference to Linnaeus (1766) in Solander's hand.

56.(**1**:52) *Squalus lebruni* (Vaillant, 1888) Squalidae

DRAWING: unfinished water-colour; *r.* [pencil] 'Sq. acanthias ?/ [ink] S. Parkinson'; *v.* [pencil] '17 Squalus maculatus/ [ink] Aehie no Mauwe'. 292 × 467.

MANUSCRIPT: Solander – [none]. Dryander – Catalogue f. 121 as sketch with colours S. Acanthias ? – N.Z., S. Parkinson.

NOTES: the verso of this drawing has an unlabelled pencil outline of *Parapercis colias* (Schneider, 1801), which is cancelled by two oblique lines. The name *Acanthias maculatus* was published by Richardson (1843*a*, 1843*b*) attributed to *Squalus maculatus* Parkinson, fig. pict. 1, f. 52 (with slight variation between the two accounts). Fortunately, this name is preoccupied by *Squalus maculatus* Bloch & Schneider, 1801, otherwise it would predate *Squalus lebruni* Vaillant, 1888.

57.(**1**:53) *Cephaloscyllium isabella* (Bonnaterre, 1788) Scyliorhinidae

DRAWING: finished water-colour, dorsal view; *r.* [pencil] 'Isabella Broussonet/

[ink] S. Parkinson'; *v.* [pencil] '8 Squalus lima/[ink] Aehie no Mauwe'. 480 × 295.

MANUSCRIPT: Solander – (D. & W. 45) S.C. Amphibia 1 ff. 189–192, 'Habitat in Oceano Novam Zelandiam alluente . . . (Nov. 24. 1769)'; (D. & W. 42) C.S.D. f. 167 (199), same details. Dryander – Catalogue f. 121, probably third entry from bottom, sketch with colours Squalus —— N.Z., S. Parkinson.

NOTES: this drawing was examined by Broussonet (1780) in the 'collection de M. le Chevalier Banks', and Broussonet's description was compiled from manuscript notes provided by Solander. Broussonet, however, provided no binominal names for his account of the dogfishes, merely referring to this species as l'Isabelle'. The name was latinized as *Squalus isabella* by Bonnaterre (1788), and independently by Gmelin (1789), both of whom relied directly on Broussonet's description.

The Solander name *lima* was published by Richardson (1843*a*, 1843*b*) as *Scyllium ? lima*, with references to Parkinson's drawing and to Broussonet.

The drawing was reproduced by Whitehead (1968).

58.(**1**:54) *Carcharhinus* sp. Carcharhinidae

DRAWING: finished pencil and wash, viewed obliquely from in front, detail of underside of head in pencil; *r.* [ink] 'Squalus Carchadius'; v. [none]. 294 × 472.

MANUSCRIPT: see no. 55. Dryander – Catalogue f. 121 as finished without colour S. Carcharias L. a tergo —— Oc., S. Parkinson.

NOTES: see no. 55.

59.(**1**:55) *Carcharhinus* sp. Carcharhinidae

DRAWING: finished pencil side view of shark, detail of underside of head by H. D. Spöring; *r.* [pencil] 'SQUALUS/Vulpecula./glaucus. vid. Phil. Trans. vol. 68 p789. fig./long. 3½ pedum.'; *v.* [ink] 'Botany Bay'. 358 × 525.

MANUSCRIPT: Solander – (D. & W. 42) C.S.D. f. 157 (189), diagnosis only; (D. & W. 40) P.N.H. f. 4 (86) 2 May 1770. Dryander – Catalogue f. 121 probably the drawing referred to as finished without colour, Squalus —— N.C., Spöring.

NOTES: the reference in the annotations to the *Philosophical Transactions* relates to the paper by Watson (1779) in which he describes a stuffed blue shark (*Prionace glauca* (L., 1758)), then in the British Museum, but caught in shallow water on the coast of Devonshire. This is also the origin of the name :'glauca' in the annotation. This drawing is not referable to *Prionace glauca*.

The *Endeavour* was at Botany Bay on 2 May 1770 (Groves, 1962).

60.(**1**:56) *Hemiscyllium ocellatum* (Bonnaterre, 1788) Hemiscyllidae

DRAWING: finished pencil side view and underside of head by H. D. Spöring; *r.* [pencil] 'SQUALUS oculatus./Sporing [partly trimmed off]'; *v.* 'Endeavours river/Hemiscyllium ocellatum Muller und Henle'. 358 × 525.

MANUSCRIPT: Solander – (D. & W. 40) P.N.H. f. 7 (89), as Squalus oculatus,

Australian Seas, Endeavour River, Careening place, 22 June 1770; length 2 feet 7 inches. Dryander – Catalogue [not listed].

NOTES: this small shark was first described by Broussonet (1784) in non-binominal form as 'l'oeillé' from a specimen two and a half feet long in the collection of Joseph Banks. He reported that 'elle a été pêchée au mois de Juillet, dans la mer du Sud, sur le côte de la nouvelle Hollande', which suggests either an error in transcription from Solander's manuscript, or the capture of another specimen which was unrecorded by Solander. Broussonet's account formed the basis for Bonnaterre's (1788) proposal of a formal name for the taxon. Independently, Gmelin (1789) also proposed the name *Squalus ocellatus* for this taxon, deriving his information from Broussonet.

Müller & Henle examined this drawing (the annotation *Hemiscyllium ocellatum* Muller und Henle is believed to have been written at the time of their visit because of the use of 'und'). They refer to a specimen of 2½ feet length in the British Museum collection at the time of their visit (before 1840). This strongly suggests that it was Banks's specimen which Broussonet described at the same length, and which in Solander's manuscript notes was described as '2 ped 7 unc' long.

Shaw (1793) described this shark, citing both Broussonet (1784) and Gmelin (1789), as 'a native of the Southern Seas', his illustration (pl. 161) was the first published of the species, and appears to have been made from a specimen, not copied from Spöring's drawing, as there are features in the engraving not visible on the drawing. This suggests that Shaw's illustration was made from the *Endeavour* specimen either in the British Museum or still in Banks's collection (the former seeming more probable). The presence of unnatural ridges by the branchial openings also suggests that the specimen was dry or stuffed. Three unlocalized stuffed specimens (f, g–h) of this species were in the British Museum collection in the late nineteenth century (Günther, 1870) but were described as half-grown, and were thus probably smaller than thirty inches in length, only one of these (specimen f) can now be found but is too small to be the *Endeavour* specimen.

A specimen from Broussonet's collection is still preserved in the Muséum National d'Histoire Naturelle, Paris (Bauchot, 1969) (MNHN – 1003); it is 355 mm in length. Although Bauchot explicitly and correctly stated it was not a type as it was not mentioned in the original description, Dingerkuss & De Fino (1983) have listed it as the holotype.

The drawing was reproduced by Whitehead (1968).

— (**1**:57) *Naso unicornis* (Forsskål, 1775) Acanthuridae

A drawing of *Naso unicornis* (Forsskål, 1775) by George Forster, labelled *r.* 'Chaetodon unicornis Anmi saw where', *v.* 'No 66 Balistoides Rhinoceros Otahite' was misbound in the *Endeavour* voyage drawings although it was clearly from the *Resolution* voyage. It was not listed by Whitehead (1978*b*) although the notes that he gives from the manuscript 'Catalogue of Forster drawings ...' (his Cat. B) probably refer to this drawing and this species not to Forster f.194 *Harpurus monoceros*. In the bound volume of the *Endeavour* drawings this folio had been numbered 57.

61.(**1**:58*a*) *Rhinecanthus rectangulus* (Bloch & Schneider, 1801) Balistidae

DRAWING: unfinished water-colour; *r.* [pencil] 'Balist. aculeatus/Ourèe Aăá/[ink] S. Parkinson'; *v.* [pencil] 'N°. 41. Balistes angulatus/[ink] Otahite'. 254 × 318.

MANUSCRIPT: Solander – (D. & W. 40) P.A.O.P. f.57 (177) Balistes angulatus Fig. Pict. (Balistes aculeatus is described at f.41 (159)). Dryander – Catalogue f.123 as Balistes aculeatus L. —— Ot., S. Parkinson sketch with colours.

NOTES: the name Balistes angulatus was published in the synonymy of *B. rectangulus* by Richardson (1848) who quoted Solander's description and referred to Parkinson's drawing.

62.(**1**:58*b*) *Naso lituratus* (Bloch & Schneider, 1801) Acanthuridae

DRAWING: unfinished water-colour; *r.* [ink] 'S. Parkinson/[pencil] Chaetodon harpuri mss varietas minor./Eoomai'; *v.* [pencil] 'N°. 31 Balistoides olivaceus/a fish of this sort [indecipherable] the same, the yellow upon it inclining to green the blue on the back broader done with/Verditer the orange spots the breadth of the tail & very bright, a circle of white after the green of the/tail the P.A. was green near the body then brown then orange the black edged w[t] white lips orange/[ink] Otahite'. 224 × 295.

MANUSCRIPT: Solander – (D. & W. 40) P.A.O.P. f.42 (160) as Balistoides olivaceus. Dryander – Catalogue f.145 as sketch with colours Chaetodon Harpurus var. Brouss. —— Soc. Isl., S. Parkinson.

NOTES: the name *olivaceus*, in the combination *Acanthurus olivacei* was used for another surgeon-fish from the *Resolution* voyage, described by Forster from Tahiti (Bloch & Schneider, 1801), but it was independently derived. The annotation Chaetodon harpuri mss varietas minor is presumed to have been written by Broussonet, or following his notes, and confirms the ascription in Dryander's Catalogue. Broussonet appears to have made a study of the 'Chaetodon' species (see Gmelin, 1789:1269).

63.(**1**:59) *Rhinecanthus aculeatus* (Linnaeus, 1758) Balistidae

DRAWING: finished water-colour; *r.* [pencil] 'Balist. aculeatus L./öidé/Oelh/oiwe tea/[ink] S. Parkinson'; *v.* [pencil] 'The colours on the back soften'd in the Orange & purple bright/N°. 50 Balistes ornatus/[ink] Otahite.' 267 × 368.

MANUSCRIPT: Solander – (D. & W. 40) P.A.O.P. f.93 (213) as Balistes ornatus. Dryander – Catalogue f.123 probably one of the two drawings listed as sketch with colours B. aculeatus L. —— Ot., S. Parkinson.

NOTES: Solander's name Balistes ornatus was published by Richardson (1848) in the synonymy of his account of *Balistes aculeatus*. Richardson quoted Solander's description verbatim.

64.(**1**:60) *Melichthys vidua* (Richardson, 1845) Balistidae

DRAWING: unfinished water-colour; *r.* [pencil] 'Aéheè tua/[ink] S. Parkinson'; *v.* [pencil] 'the whole body fusca black towards the tail & paler on the face & breast/ the iris of the eye olive pupil black./N°. 29 Balistes vidua/[ink] Otahite'. 267 × 371.

MANUSCRIPT: Solander – (D. & W. 40) P.A.O.P. f.36 (154) as Balistes vidua. Dryander – Catalogue f.123 as sketch with colours Balistes vidua Brouss. —— Ot., S. Parkinson.

NOTES: Richardson (1845) used Solander's name *Balistes vidua* properly attributing it to Solander's authorship and referring to this drawing as 'Parkinson in Bibl. Banks 1. No.60. 29'. He also cited George Forster's drawing ('Forster in Bibl. Banks. No.246') also from Tahiti. There is a specimen in the British Museum (Natural History) 1972.8.16.1. which Günther (1870) described as 'Adult: bleached. Otaheiti. Old Collection. Probably from Cook's voyage, and type of the Species' which if strictly interpreted should mean it was an *Endeavour* specimen. This specimen is 165 mm standard length (192 mm t.l.). Unfortunately, Solander gave no measurements for the specimens he examined and he apparently preserved seven specimens of the species. There is thus no way of telling whether this specimen is an *Endeavour* fish, or whether it was the one drawn – the drawing measures 152 mm S.L. (182 mm t.l.) and is thus close to the specimen in length. It could equally well be a *Resolution* specimen drawn by Forster and approaches that drawing closely in size; Wheeler (1981) considered it more likely to be the Forster specimen. It still possesses type status.

65.(**1**:61) *Balistoides viridescens* (Bloch & Schneider, 1801) Balistidae

DRAWING: unfinished water-colour; *r.* [pencil] 'Oiree Nemoo/[ink] S. Parkinson'; *v.* [pencil] 'N°. 57. Balistes Gigas/[ink] Otahite'. 266 × 371.

MANUSCRIPT: Solander – (D. & W. 40) P.A.O.P. f.65 (185) as Balistes gigas, Fig. Pict. Dryander – Catalogue probably f.123 sketch with colours, Balistes —— Ot., S. Parkinson, one of two entries so labelled.

NOTES: Solander's manuscript name does not appear to have been used by subsequent authors.

66.(**1**:62) *Cantherhines dumerili* (Hollard, 1854) Balistidae

DRAWING: unfinished water-colour; *r.* [pencil] 'Oilhe roweppa/[ink] S. Parkinson'; *v.* [pencil] 'The whole body of the fish fusca. the tail the same lighter at the tip the part/at the beginning of the tail dirty white the finns transparent the ribs/dirty yellow the Iris of the eye orange pupil black./72. Balistes chrysopterus/[ink] Otahite'. 267 × 370.

MANUSCRIPT: Solander – (D. & W. 40) P.A.O.P. f.129 (249) as Balistes chrysopterus. Dryander – Catalogue probably f.123 sketch with colours, Balistes —— Ot., S. Parkinson, one of two entries so labelled.

NOTES: Schneider's (1801) use of the name *Balistes chrysopterus* appears to have been independent of Solander's manuscript.

67.(**1**:63) *Arothron meleagris* (Bloch & Schneider, 1801) Tetraodontidae

DRAWING: pencil with some colour; *r*. [pencil] 'Ehooi/[ink] S. Parkinson'; *v*. [pencil] 'the whole of this Fish fins & all is a purple black spotted with milk colour'd spots/the teeth dirty white/N°. 49 Tetraodon Meleagris/[ink] Otahite'. 270 × 371.

MANUSCRIPT: Solander – (D. & W. 40) P.A.O.P. f.78 (198) as Tetraodon meleagris; a note records that the specimen was skinned. Dryander – Catalogue f.125, one of two listed as sketch with colours, Tetrodon —— Ot., S. Parkinson.

NOTES: Schneider's (1801) name *Tetrodon meleagris* was derived from Lacepède's (1798) use of Le Tetrodon méléagris, the details of which in turn derived from a manuscript of Commerson. These names appear to be independent of Solander's name.

68.(**1**:64) *Aluterus scriptus* (Osbeck, 1765) Balistidae

DRAWING: finished water-colour; *r*. [ink] 'Balistes monoceros./Sydney Parkinson pinx^{t} 1768'; *v*. [ink] 'Octr. 4 1768/Lat N.'. 234 × 292.

Fig. 6 *Aluterus scriptus* (Osbeck, 1765). A Parkinson drawing made in mid-Atlantic on 4 October 1768. (Catalogue number 68.)

MANUSCRIPT: Solander – (D. & W. 42) C.S.D. f.191 (133) Oceano Atlantico inter tropicos Lat N. 11°, Oct. 4. 1768 captus as Balistes Monoceros; (D. & W. 45) S.C. Amphibia 1, f.238–239 – same data. Dryander – Catalogue f.123 as finished in colour Balistes Monoceros L. —— Oc., S. Parkinson.

NOTES: the name *Balistes monoceros* was taken from Linnaeus (1766), as the citation in the Solander Slip Catalogue makes clear (although the name dates from Linnaeus, 1758).

69.(**1**:65) *Lagocephalus spadiceus* (Richardson, 1845) Tetraodontidae

DRAWING: pencil; *r.* [ink] 'S. Parkinson'; *v.* [pencil] 'The black dark olive grey fading on the sides into brassy colour the belly white the P.D. &/upper part of the tail brassy olive the other fins white the iris yellow pupil black./Tetraodon assellinus/[ink] Endeavour's River'. 267 × 367.

MANUSCRIPT: Solander – (D. & W. 40) P.N.H. ff.12–15 (94) Habitat near Endeavour River careening place. Dryander – Catalogue f.125, probably the drawing entered as sketch without colours, Tetrodon —— N.C., S. Parkinson.

NOTES: the name Tetraodon assellinus, attributed to Solander MS *Pisces Novae Hollandiae*, was published by Richardson (1845) in the synonymy of his *T. spadiceus*. Richardson also referred to Parkinson's drawing (no.65) in the Banksian library.

70.(**1**:66) *Canthigaster solandri* (Richardson, 1845) Tetraodontidae

DRAWING: unfinished water-colour; *r.* [pencil] 'Tāitāi/[ink] S. Parkinson'. *v.* [pencil] 'every spot is border'd w[t] a dark line which turns paler as the/ground colour does./N°. 28 Tetraodon punctatus/[ink] Otahite'. 295 × 284.

MANUSCRIPT: Solander – (D. & W. 40) P.A.O.P. f.29 (147). Dryander – Catalogue f.125 as sketch with colours, Tetrodon —— Ot., S. Parkinson.

NOTES: Richardson (1845) cited this drawing in the Banksian library quoting both the folio number of the bound volume (66) and that of an earlier sequence 56, incorrectly for 57. He also incorrectly quoted the drawing as being labelled *Tetrodon cinctus* (although he did not attribute the name to Solander but suggested it was a later addition). In fact the name *T. cinctus* is written on the lower of the two drawings when they were mounted on a sheet for binding, and refers to no.71 of this catalogue. Richardson had a specimen from Belcher's voyage on the *Sulphur*, which he described and which is still extant. Later, Richardson (1848) corrected his attribution of the name *T. cinctus* to this drawing, correctly referred to it as *T. punctatus*, and quoted Solander's description of it in extenso. This followed re-examination of the Solander manuscripts and Parkinson drawings.

71.(**1**:66) *Arothron stellatus* (Bloch & Schneider, 1801) Tetraodontidae

DRAWING: unfinished water-colour; *r.* [pencil] 'Hue Hue eti/Taitai/[ink] S. Parkinson/Tetraod. cinctus Specis'; *v.* [pencil] 'N°. 56/[ink] Otahite'. 236 × 284.

MANUSCRIPT: Solander – not identifiable in any mss. Dryander – Catalogue f. 125 as sketch with colours T. cinctus Brouss. —— Ot., S. Parkinson.

NOTES: the name Tetrodon cinctus was incorrectly attributed by Richardson (1845) – see discussion above – but was published with a description derived from the Parkinson drawing by Richardson (1848). He was clearly uncertain of its identity, and suggested that it closely resembled *T. lineatus* 'of the *Fauna Japonica*'. Richardson could not trace any description of this fish in the Solander manuscripts; if the Dryander Catalogue attribution of the name T. cinctus to P. M. A. Broussonet, is correct, then this would account for the name not appearing in Solander's manuscripts. Like other names, notably those in the genus *Chaetodon*, this was a Broussonet manuscript name.

72. (**1**:67) *Chilomycterus* cf. *antillarum* Jordan & Rutter, 1897 Diodontidae

DRAWING: partly coloured pencil sketch; *r.* [ink] 'S. Parkinson'; *v.* [pencil] 'the whole body & spines an olive colour the fins yellow – the spots mark'd & are black/ the pupil of the eye black the iris gold colour gradually fading into the olive &/very prominent/N°. 2 Diodon/[ink] Brasil'. 267 × 351.

Fig. 7 *Canthigaster solandri* (Richardson, 1845). Parkinson's drawing made at Tahiti which was later used in part by Richardson (1845) to name this sharp-nosed puffer-fish in honour of Daniel Solander. (Catalogue number 70.)

MANUSCRIPT: Solander – not identifiable in any mss. Dryander – Catalogue f. 127 as sketch with colours Diodon —— Bras., S. Parkinson.

NOTES: although there are three species of *Diodon* described in Solander's Slip Catalogue, Amphibia 1 (D. & W. 45) *viz.* D. erinaceus (f.261), D. aculeatus (f.262), and D. truncatus (f.263 – originally described as *Balistes*) none can be related to this Brazilian fish. The first relates to the next drawing (no.73, this catalogue), the other two were both from Jamaica and not *Endeavour* specimens. The identification of this drawing is tentative; too few features are clearly shown for certain identification.

73.(**1**:68) *Diodon hystrix* Linnaeus, 1758 Diodontidae

DRAWING: finished water-colour, two views the second of the fish inflated; *r.* [ink] 'Diodon Erinaceus/Sydney Parkinson pinx[t] ad vivum 1768'; *v.* [ink] 'Oct[r]. 7th. 1768/[ink] Lat. N.' 235 × 292.

MANUSCRIPT: Solander – (D. & W. 45) S.C. Amphibia 1 f.261 as Diodon Erinaceus (this trivial name replaces Hystrix as originally written) 'Habitat in Oceano Atlantico. Lat. Sept IX.43. Oct. 7. 1768'. Dryander – Catalogue f. 127 as finished in colours D. Erinaceus mss – Oc., S. Parkinson.

NOTES: Solander's name Diodon erinaceus seems not to have been published although Agassiz (1841) independently used this combination for a fossil fish.

This *Endeavour* specimen, captured in mid-Atlantic between the landfalls of Madeira and Rio de Janeiro, was one inch (26 mm) long, and exhibits the body form and colouring typical of the pelagic juvenile phase (Leis, 1977).

See also no.229 in this catalogue for a preliminary sketch.

74.(**2**:1) *Muraena helena* Linnaeus, 1758 Muraenidae

DRAWING: finished water-colour; *r.* [ink] 'Muraena guttata/Sydney Parkinson pinx[t] 1768'; *v.* [ink] 'Madeira/Rio Janeiro'. 299 × 480.

MANUSCRIPT: Solander – (D. & W. 45) S.C. Pisces 1, ff. 8–9 as Muraena guttata, 'Habitat ad Insulam Maderam Oceani Atlantici, etjam in Portu Fluvii St Januarii in Brasilia'; (D. & W. 42) C.S.D. f.201 (216) with same data, one of the two specimens was thirty inches long. Dryander – Catalogue f. 131 as finished in colour M. guttata Mss —— Madeira, S. Parkinson.

NOTES: the Solander name *Muraena guttata* was published by Richardson (1848) who referred to both the Parkinson drawing and to a Banks–Solander manuscript. Unfortunately, Solander's name is not available for use as at the date of its publication by Richardson it was preoccupied nomenclaturally by *Muraena guttata* Risso, 1826. Kaup (1856) used the Solander name from Richardson as the type species of his genus *Limamuraena* but referred to Solander's manuscript again for details of the conformation of the nostrils; he specifically rejected Risso's name.

This drawing appears to have been made at Madeira as was Solander's manuscript description. It seems that another moray eel was obtained at Rio de Janeiro, which was thought to be the same species, and reference to this caused the confusion in locality ascriptions noted in Solander's notes.

75.(**2**:2) *Gymnothorax ocellatus* Agassiz, 1831 Muraenidae

DRAWING: finished water-colour; *r.* [pencil] 'muraena tricolor mss in mus. britt/ [ink] S. Parkinson'; *v.* [pencil] '15 Muraena/[ink] Brasil'. 273 × 356.

MANUSCRIPT: Solander – none (see under Notes). Dryander – Catalogue f. 131 as finished in colour M. tricolor Brouss. —— Brasil, S. Parkinson.

NOTES: the identification of this drawing is not certain and it may represent the species known as *G. nigromarginatus* Girard, 1858. No question of nomenclatural priority is involved, however, as Muraena tricolor is an unpublished name. Its attribution to Broussonet by Dryander is interesting and suggests that this may be the Brazilian fish which Solander considered to be identical with his Madeiran *Muraena guttata*; this would account for the absence of a separate entry in his manuscript notes. Despite Broussonet's annotation of 'in mus. britt.' no specimen which could be associated with this Brazilian fish is listed by Günther (1870) nor can be located now.

76.(**2**:3) *Echidna nebulosa* (Ahl, 1789) Muraenidae

DRAWING: unfinished water-colour; *r.* [ink] 'S. Parkinson/[pencil] pepedho'; *v.* [pencil] '69. Muraena geographica'. 298 × 470.

MANUSCRIPT: Solander – (D. & W. 40c) P.A.O.P. f. 132 (252) as Muraena geographica; vernacular names; Habitat Ulhaietea. Dryander – Catalogue f. 131 as sketch with colours Muraena —— Ulietea.

NOTES: this eel was collected at Uhlietea (= Raiatea) in the Society Islands between 20 and 24 July 1769; it was preserved in Cagg No 6, but no specimen in the Museum's collection was recognizable as the *Endeavour* specimen in Günther's *Catalogue of Fishes* (1870). The Solander manuscript description and Parkinson's figure were referred to and discussed by Richardson (1844–1848) and in almost identical phraseology by Kaup (1856). The name Muraena geographica was published in synonymy by Richardson.

77.(**2**:4) *Istiblennius lineatus* (Valenciennes in Cuvier & Valenciennes, 1836) Blenniidae

DRAWING: unfinished water-colour; *r.* [ink] 'S. Parkinson/[pencil] Ohooa'; *v.* [pencil] 'N° 46. Blennius lineatus/[ink] Otahite'. 269 × 370.

MANUSCRIPT: Solander – (D. & W. 40c) P.A.O.P. f. 79 (199) as Blennius lineatus; vernacular names. Dryander – Catalogue f. 137 as sketch with colours, Blennius —— Otahite, S. Parkinson.

NOTES: the use of the name *lineatus* by Valenciennes appears to have been independent from that of Solander. Although a specimen was preserved in Cagg No. 5 it could not be traced in the mid-1800s in the collection of the British Museum (Günther, 1861).

78.(**2**:5) *Pseudophycis bachus* (Forster in Bloch & Schneider, 1801) Moridae

DRAWING: unfinished water-colour; *r.* [ink] 'S. Parkinson'; *v.* [pencil] '18. Blennius venustus/ [ink] Totarranue'. 293 × 460.

MANUSCRIPT: Solander – (D. & W. 40a) P.A. index (f.63) listed only. Dryander – Catalogue f.137 as sketch with colours, *Blennius* —— New Zealand, S. Parkinson.

NOTES: Solander's name venustus was published by Richardson (1843b) as *Brosmius venustus*. Richardson cited the Parkinson drawing but commented that he could not find the species described by Solander although he suggested that Solander's Blennius rubiginosus at f.14 (16) of *Pisces Australiae* (D. & W. 40a) might refer to it. This seems unlikely and although there is no description in the Solander manuscript, Blennius venustus is listed in the index to the fishes from New Zealand at folio 63. This index shows that one specimen only was obtained, gives no folio reference to its description and no serial number was allocated to it (this suggests that it was not preserved).

This drawing was reproduced by Whitehead (1968).

79.(**2**:6) *Echeneis neucrates* Linnaeus, 1758 Echeneidae

DRAWING: unfinished water-colour; *r.* [ink] 'S. Parkinson/ [pencil] Echeneis neucrates L.'; *v.* [pencil] 'N°. 6/ [ink] Otahite'. 298 × 472.

MANUSCRIPT: Solander – (D. & W. 40c) P.A.O.P. f.113 (233) as Echeneis anguillaris from Tahiti, serial number A.176, Cagg No.6, one specimen; (D. & W. 40b) P.N.H. f.5 (87) as Echeneis neucrates Australian seas 24 May 1770. Dryander – Catalogue f.139 as sketch with colours, E. Neucrates L. Otahiti, S. Parkinson.

NOTES: the name *Echeneis neucrates* on the drawing appears to be an early identification which was later adopted by Dryander who also wrote the locality Otahite. If Dryander was correct in attributing the drawing to this locality then the specimen was the one which Solander described as Echeneis anguillaris; conversely, if the identification was correct, or written by Solander, then this specimen would have been the *E. neucrates* from Australia caught on 24 May 1770. At this date the *Endeavour* was at Bustard Bay, Queensland, 24°05'S., 151°48'E. (Groves, 1962)

The Solander name Echeneis anguillaris does not seem to have been published by later authors.

80.(**2**:7) *Coryphaena hippurus* Linnaeus, 1758 Corpyhaenidae

DRAWING: pencil sketch; *r.* [ink] 'S. Parkinson/ [pencil] 15½'; *v.* [pencil] 'The predominent colours of this fish are blue, green & yellow the blue occupies/the back fin & parts of the back this gradually turns to Green on the sides & that/to a fine yellow which takes place on the belly & fins besides it is all over spotted w^t blue/the iris of the eye yellow green. – /N°. 4. Coryphaena Hippuris/ [ink] Rio de Janeiro'. 298 × 480.

MANUSCRIPT: Solander – (D. & W. 45) S.C. Pisces 1, f.99, and (D. & W. 42)

C.S.D. f.209 (204), both giving two sets of meristic data and measurements of a specimen of 39 inches total length. Dryander – Catalogue f.139 as sketch without colours, *Coryphaena Hippuris* L. —— Brasil, S. Parkinson.

NOTES: the dolphin-fish was a well-known species which was presumably the reason Solander did not describe it fully. Neither of the two sets of meristic data are identical with those in Linnaeus (1766). Therefore it can be assumed that two specimens were examined. However, there is only the one set of measurements. As the specimens captured in the Atlantic Ocean were recorded in the Slip Catalogue only (and the foolscap manuscript Copies of Solander's descriptions was compiled from them) it seems very probable that the measurements of the 39-inch fish actually refer to the specimen drawn.

81.(**2**:8) *Xyrichthys novacula* (Linnaeus, 1758) Labridae

DRAWING: finished water-colour; *r.* [ink] 'Coryphaena Novacula./Sydney Parkinson pinx[t] 1768/I.19. Madeira/[pencil] Coryphaena Novacula'; *v.* [pencil sketch of head of eel]/[pencil] 'Mem. the M.B. goes no further than opposite to the eye' [this probably refers to the branchiostegal membranes of the eel]. 238 × 295.

MANUSCRIPT: Solander – (D. & W. 45) S.C. Pisces 1, f.107, and (D. & W. 42) C.S.D. f.211 (206); both with a full description of a fish six inches long from Madeira, vernacular name Papagaya. Dryander – Catalogue f.139 as finished in colours, *C. Novacula* L. —— Maderia, S. Parkinson.

NOTES: this is one of the finished drawings by Parkinson typical of his work at Madeira.

82.(**2**:9) *Xyrichthys pentadactylus* (Linnaeus, 1758) Labridae

DRAWING: unfinished water-colour; *r.* [ink] 'S. Parkinson/[pencil] Paou'; *v.* [pencil] 'N°. 27. Coryphaena virens/[ink] Otahite'. 271 × 370.

MANUSCRIPT: Solander – (D. & W. 40c) P.A.O.P. f.37 (155) as Coryphaena virens, vernacular name, one specimen serial number A71 in Cagg 3, and one, unmarked (= unnumbered) in Cagg 5; that there were two specimens is confirmed in the index to this manuscript (f.288). Dryander – Catalogue f.139 as sketch with colours Coryphaena —— Otahite, S. Parkinson.

NOTES: this drawing was copied for Cuvier and the name '*Coryphaena virens*' was attributed to Parkinson and published as *Xyrichthys virens* Valenciennes in Cuvier & Valenciennes, 1839. The drawing therefore has type status. The name *Coryphaena virens* Gmelin, 1789 appears to be a proposal of the name independent of Solander's usage but may have been another example of Broussonet communicating information from Banks's collections to Gmelin.

Although two specimens were preserved on the *Endeavour* neither seems to have been recognizable when Günther (1862) listed these wrasses in the British Museum.

83.(**2**:10) *Valenciennea strigatus* (Broussonet, 1782) Gobiidae

DRAWING: unfinished water-colour; *r.* [ink] 'S. Parkinson/[pencil] Gobius strigatus Broussonet Ichth./Teipooa'; *v.* [pencil] 'N°. 60 Labrus delicatulus/[ink] Otahite'. 270 × 370.

MANUSCRIPT: Solander – (D. & W. 40c) P.A.O.P. f.71 (191) as Labrus delicatulus, vernacular name, one specimen serial number A129 in Cagg number 1. Dryander – Catalogue f.139 as sketch with colours Gobius strigatus Brouss. —— Otaheite, S. Parkinson.

NOTES: Broussonet (1782) named this species *Gobius strigatus* in part on the Parkinson drawing and on Solander's manuscript description (the vernacular name Taiboa given in the Solander manuscript is identical in Broussonet, whereas the spelling on the drawing differs). Broussonet also cited the vernacular name Taipoa attributing it to J. R. Forster, and the locality he gave '*Oceanus* pacificus prope Insulam Otaheite, *J. R. Forster* (Mus. Britannic.)' implies that he employed a specimen from the *Resolution* voyage, collected by J. R. Forster, in the British Museum, for his description. There is a drawing (f.189) in the Forster drawings (Wheeler, 1981), but the species is not described in the published account of the zoology of the voyage (Forster, 1844). No specimens from either voyage exist in the British Museum (Natural History).

The Parkinson drawing appears to be the original for Nodder's illustration reproduced in Broussonet (1782).

84.(**2**:11) *Platycephalus fuscus* Cuvier & Valenciennes, 1829 Platycephalidae

DRAWING: unfinished water-colour; *r.* [ink] 'S. Parkinson/[pencil] Earai era-ere.'; *v.* [pencil] 'N° 63. Cottus Otahitensis./[ink] Otahite'. 272 × 374.

MANUSCRIPT: Solander – (D. & W. 40c) P.A.O.P. f.114 (234) as Cottus otaheitensis, vernacular names, one specimen serial number A177 in Cagg number 6. Dryander – Catalogue f.141 as sketch with colour, Cottus —— Otahite, S. Parkinson.

NOTES: this drawing was mentioned by Cuvier (1829) in the description of *P. fuscus*, and the name *Cottus otaitensis* attributed to Parkinson; Cuvier also quoted from Solander's manuscript description in particular notes on the coloration and the vernacular names.

85.(**2**:12) *Scorpaena cardinalis* Richardson, 1842 Scorpaenidae

DRAWING: pencil; *r.* [ink] 'S. Parkinson'; *v.* [pencil] 'the body the collour of minium spotted w[t] dark red, mark'd on the belly w[t] white the head scarlet spotted/ above with dark red & below w[t] white the finns rather yellower than the body & mark'd with/dark red also the PA yellow spotted w[t] red the end of the tail yellow/ 10. Scorpaena Cardinalis/[ink] aehie no Mauwe.' [pencil sketch of a crawfish.] 292 × 461.

MANUSCRIPT: Solander – (D. & W. 40a) P.A. f.36 (38) as Scorpaena cardinalis, habitat off Motuaro. Dryander – Catalogue f.141 as sketch without colours, Scorpaena —— New Zealand, S. Parkinson.

NOTES: this drawing and Solander's manuscript were cited by Richardson (1842*a*) as the sole source of information for his species. He quoted Solander's description almost verbatim. He also (Richardson, 1843*b*) referred to the differences between the locality given on the drawing, 'Eaheenomauwee' (= North Island of New Zealand) and in Solander's manuscript as 'Motuaro' (in the Bay of Islands). Richardson (1842*a*) also compared this drawing with Forster's drawing (vol. 2, f.190) of Scorpaena Cottoides (see Wheeler, 1981) of a specimen from Dusky Bay, but considered the latter to be a distinct species.

This drawing was reproduced by Whitehead (1968).

86.(**2**:13) *Scorpaenopsis gibbosa* Bloch & Schneider, 1801 Scorpaenidae

DRAWING: unfinished pencil sketch; *r.* [pencil] 'Scorpaena diabolus/Nohoo-/Noohoo-noohoa teraou/[ink] S. Parkinson'; *v.* [pencil] 'N°. 52. Scorpaena marmorata/[ink] Otahite.' 271 × 371.

MANUSCRIPT: Solander – (D. & W. 40c) P.A.O.P. f.75, (195) as Scorpaena marmorata. Dryander – Catalogue f.141 as Scorpaena —— sketch without colour Otaheite, S. Parkinson.

NOTES: Solander's notes in P.A.O.P. show that there were two specimens of Scorpaena marmorata, both were preserved: one numbered 122 in Cagg 5, the other numbered 187 in Cagg 6. The Parkinson drawing was referred to by Cuvier (in Cuvier & Valenciennes, 1829) under the name *Scorpaena diabolus*, but was not employed in Cuvier's diagnosis of the species. The figure was also cited by Richardson (1844–1845) but he apparently did not refer to the Solander description.

The drawing was reproduced by Whitehead (1968) as plate 32 with the caption to the drawing misplaced and labelled Plate 33; the annotations on the drawing were reproduced incorrectly.

The vernacular name given on the drawing is similar to the 'mohu tarao' given for *Scorpaenopsis* sp. by Randall (1973) and Bagnis *et al.* (1972).

No specimen which could have been *Endeavour* material was listed by Günther (1860).

87.(**2**:14) *Pterois radiata* Cuvier in Cuvier & Valenciennes, 1829 Scorpaenidae

DRAWING: unfinished water-colour; *r.* [ink] 'Pterois radiata/[pencil] gasterosteus volitans Linn. Scorpaena/[pencil] Tatary/Tataraìhíāu/[ink] S. Parkinson'; *v.* [pencil] 'N°. 41 Scorpaena radiata/[ink] Otaheite'. 272 × 373.

MANUSCRIPT: Solander – (D. & W. 40c) P.A.O.P. f.38 (156) as Scorpaena radiata – specimen A76 preserved in Cagg No.3. Dryander – Catalogue f.141 as Scorpaena —— sketch with colours, Otahite, S. Parkinson.

NOTES: Cuvier (1829) described *Pterois radiata* solely on the basis of the drawing by Parkinson, but incorrectly alleged that Solander's notes contained nothing pertaining to the drawing. The Parkinson drawing is the sole source for the taxon and has type status. This drawing was reproduced by Whitehead (1968) as Plate 34.

88.(**2**:15) *Scorpaena porcus* Linnaeus, 1758 Scorpaenidae

DRAWING: finished water-colour; *r.* [ink] 'Scorpaena Patriarcha./Sydney Parkinson. pinx[t] 1768'; *v.* [ink] 'Madera'. 293 × 462.

MANUSCRIPT: Solander – (D. & W. 42) C.S.D. f.217 (212) as Scorpaena patriarcha, habitat in Oceano Atlantico ad insulam Maderam, 14½ unc; (D. & W. 45) S.C. Pisces 1, ff.130–133v, same details. Dryander – Catalogue f.141 as Scorpaena Patriarcha, finished in colours —— Madeira, S. Parkinson.

NOTES: Solander's description in the Slip Catalogue are the original notes, the others are a later copy. The description is very detailed and Solander assumed he was dealing with a previously unknown species having failed to recognize his specimen as Linnaeus's *Scorpaena porcus*. Substantial parts of the description have been altered, as have those of his Scorpaena choirista (see number 91 of this catalogue); it is evident that the descriptions of both species became confused with one another.

This drawing was reproduced by Wheeler (1983) at plate 195.

89.(**2**:16) *Helicolenus papillosus* (Schneider in Bloch & Schneider, 1801) Scorpaenidae

DRAWING: unfinished water-colour; *r.* [ink] 'S. Parkinson'; *v.* [pencil] 'the spots on the linea lateralis white/14. Scorpaena percoides/[ink] Motuaro'. 298 × 476.

MANUSCRIPT: Solander – (D. & W. 40a) P.A. f.4 (6) as Scorpaena percoides, habitat Cape Kidnappers. Dryander – Catalogue as Scorpaena ——, sketch with colours, New Zealand, S. Parkinson.

NOTES: the *Endeavour* was at Motuaro in the Bay of Islands, North Island of New Zealand on 29 November and 2 December 1769. Richardson (1842*a*), who referred extensively to Solander's description and this drawing, alleged that this species was captured both off Cape Kidnappers and at Motuaro in Queen Charlotte's Sound, but it seems that only one specimen was captured on this voyage. *Sebastes percoides* was named by Richardson from the Solander manuscript and the drawing which therefore has type status. The drawing was reproduced by Whitehead (1968) as Plate 33, but the caption referring to this drawing is numbered 32.

Schneider's name *Synanceja papillosus* was based solely on Forster's manuscript description of a fish from Dusky Bay caught 30 March 1773 on the *Resolution* voyage and labelled *Scorpaena cottoides*. The drawing which was made from this specimen was briefly mentioned by Richardson (1842*a*) under *Scorpaena cardinalis*, which was based on a drawing by Parkinson and Solander's description (see number 85 of this catalogue). Cuvier (1829) refers to the Forster name under his account of *Scorpaena cirrhosa*.

90.(**2**:17) *Synanceja verrucosa* Bloch & Schneider, 1801 Scorpaenidae

DRAWING: unfinished water-colour; *r.* [ink] 'S. Parkinson/[pencil] Ehoohoo pooa

pooa/D[r] Vivian'; *v.* [pencil] 'N° 65 Scorpaena horrida/The ground of the fish is rather darker cover'd all over w[t] a sort of shiny tubercles the tail & finns strip'd much in the manner of the other Scorpaena's the spots also darker/The eyes blueish grey./[ink] Otahite.' 298 × 417.

MANUSCRIPT: Solander – (D. & W. 40c) P.A.O.P. f.115 (235) as Scorpaena horrida, one fish, vernacular names, specimen no A178 in Cagg number 6. Dryander – Catalogue f.141 as sketch with colours, Scorpaena —— Otahite, S. Parkinson.

NOTES: Solander identified this fish with Linnaeus's *Scorpaena horrida*; no later worker appears to have referred to the drawing. Although the specimen was preserved it does not appear in Günther's (1860) listing of the scorpaenid fishes in the British Museum.

91.(**2**:18*a*) *Pontinus kuhlii* (Bowdich, 1825) Scorpaenidae

DRAWING: finished water-colour; *r.* [ink] 'Scorpaena Choirista./Sydney Parkinson pinx[t] 1768/T. 18. Madeira'; *v.* [pencil] 'Mem the Pinna dorsalis more diverging at front & center M.B. 7. the first spine of the/P.A. too long'. 234 × 292.

MANUSCRIPT: Solander – (D. & W. 45) S.C. Pisces 1, ff.134–137*v* as Scorpaena chorista, Habitat in Oceano Maderam alluente, length of fish 8 inches; (D. & W. 42) C.S.D. f.213 (208) same data. Dryander – Catalogue f.141, as finished in colours, Scorpaena Choirista mss —— Madeira, S. Parkinson.

NOTES: this species was undescribed at the time of its capture and was not formally named for nearly 50 years. Unfortunately Solander's detailed description was not published and his name is unavailable. The name 'chorista' is explained by reference to the Solander manuscripts where the Madeiran name is recorded as 'Menino do Cor (Boy of the Choir)' (S.C. Pisces 1, f.134) and at f.137*v* by a note recording its colours as those of the vestments of a church choirboy. The rose colour of Parkinson's drawing suggests that choirboys wore some colour other than the present pale lilac in eighteenth-century Madeira.

A specimen, listed by Günther (1860) as 'e. Adult. Old Collection, as *Scorpaena chorista*', is undoubtedly the specimen from the *Endeavour* voyage and has recently been labelled as such in the collection of the British Museum (Natural History).

92.(**2**:18*b*) *Scorpaena isthmensis* Meek & Hildebrand, 1928 Scorpaenidae

DRAWING: finished water-colour; *r.* [ink] 'S. Parkinson'; *v.* [pencil] 'N° 15 Scorpaena/[ink] Brasil'. 232 × 289.

MANUSCRIPT: Solander – none. Dryander – Catalogue f.141 as finished in colours, Scorpaena —— Brasil, S. Parkinson.

NOTES: no scorpaenid appears in Solander's manuscript notes (either S.C. or C.S.D. – D. & W. 45 and 42) on Brazilian fishes, and as the drawing bears no trivial name, it has to be assumed that this fish was not examined or described by Solander. The identification of this drawing is tentative bearing in mind the difficulty in identifying a scorpaenid fish from a drawing, but the presence of a conspicuous dark blotch on the spiny dorsal fin, the dark colouring of the posterior end of the anal fin, and the dark posterior half of the pelvic fins all indicate that this fish was *S. isthmensis*, which is recorded from shallow water off Rio de Janeiro (Eschmeyer, 1965, 1969).

93.(**2**:19*a*) *Bothus podas maderensis* (Lowe, 1834) Bothidae

DRAWING: unfinished water-colour; *r.* [ink] 'S. Parkinson'; *v.* [pencil] 'N°. 3. Pleuronectes Rhomboides/ [ink] Madera'. 235 × 286.

MANUSCRIPT: Solander – (D. & W. 45) S.C. Pisces 1, ff.156–157v as Pleuronectes Rhomboides, habitat in Mare Atlantico prope Insulam Maderam, total length 6½ inches; (D. & W. 42) C.S.D. f.219 (214) same data. Dryander – Catalogue as sketch with colours, Pleuronectes —— Brasil, S. Parkinson.

NOTES: Solander's trivial name Rhomboides was derived from the Rondelet (1554) classical name, which had been employed for the Mediterranean form of the species by many pre-Linnaean authors. There were no specimens in the British Museum which could have been the *Endeavour* material (Günther, 1862).

94.(**2**:19*b*) *Gymnachirus nudus* Kaup, 1858 Soleidae

DRAWING: unfinished water-colour; *r.* [ink] 'S. Parkinson'; *v.* [pencil] 'all round the fins & the tail is a narrow border of white/the 3 last stripes but one are paler than the rest/the last of all at the end of the tail is almost black./N°. Pleuronectes/ [ink] Brasil'. 237 × 291.

MANUSCRIPT: Solander – none. Dryander – Catalogue f.143 as sketch with colours, Pleuronectes —— Brasil, S. Parkinson.

NOTES: this seems to be yet another case of Solander failing to compile a description or propose a name for a specimen drawn by Parkinson at Brazil.

95.(**2**:20) *Bothus mancus* (Broussonet, 1782) Bothidae

DRAWING: unfinished water-colour; *r.* [pencil] 'Patee maure/Pleuronectes pictus [deleted] mancus mss/ [ink] S. Parkinson'; *v.* [pencil] '75 Pleuronectes maculata [ink] Ulhietea'. 272 × 372.

MANUSCRIPT: Solander – (D. & W. 40c) P.A.O.P. f.132 (252) as Pleuronectes maculata, vernacular name, length 15 inches. Dryander – Catalogue f.143 as sketch with colours, P. Mancus Brouss. —— Society Islands, S. Parkinson.

NOTES: *Pleuronectes mancus* was described by Broussonet (1782) from a series of literary sources, most of which derived from Marcgrave (1658) and were thus from South American (Brazilian) waters, but also referring to both Solander's manuscript, and the manuscript and a specimen from J. R. Forster's voyage on the *Resolution*. Although Broussonet did not cite the Parkinson drawing (nor that of G. Forster – folio 192) he undoubtedly referred to them, and the annotation noted above in which 'Pleuronectes pictus' was changed to 'Pleuronectes mancus mss' is probably Broussonet's. However, he certainly consulted the manuscript, and the *Endeavour* material has some claim to type status. His major source of information, however, appears to have been the Forster specimen preserved in the British Museum, collected at Anamoka (Namuka, Friendly Islands) on 28 June 1774, which was 114 lines (equals 11.4 inches) in standard length (see Wheeler, 1964). The Forster fish is no longer present in the collection of the British Museum (Natural History) (Wheeler, 1981).

The index of P.A.O.P. compiled by Solander (D. & W. 40c, folio 292) shows that the *Endeavour* specimen was not preserved.

96.(**2**:21) *Drepane punctata* (Linnaeus, 1758) Ephippidae

DRAWING: pencil sketch; *r.* [ink] 'S. Parkinson'; *v.* [pencil] 'The whole fish silvery the spots fusca/Chaetodon punctatus L./[ink] Endeavours River'. 269 × 374.

MANUSCRIPT: Solander – (D. & W. 40b) P.N.H. f.9–21 (91) as Chaetodon punctatus, habitat near Endeavour River Careening Place. C. punctatus was also recorded at Tahiti see P.A.O.P. f.15* (132). Dryander – Catalogue f.147 as sketch without colours Chaetodon —— N.C. [Nova Cambria], S. Parkinson.

NOTES: Solander correctly identified the fish from the Endeavour River, caught between 17 June and 4 August 1770 with Linnaeus's species. His description was cited by Cuvier (1831).

It is not possible to relate any specimen in the British Museum (Natural History) to the Endeavour River fish.

97.(**2**:22*a*) *Zebrasoma scopas* (Valenciennes in Cuvier & Valenciennes, 1835) Acanthuridae

DRAWING: finished water-colour; *r.* [pencil] 'Chaetodon an militaris'; *v.* [pencil] 'N°. 32 Zeus Elevatus/Erapepe/Taumata, the same name/with their [unreadable word]/[ink] Otahite'. 233 × 295.

MANUSCRIPT: Solander – (D. & W. 40c) P.A.O.P. f.45, (163) as Zeus elevatus, vernacular names, specimen number A88 in Cagg No.3; Dryander – Catalogue f.141 as sketch with colours, Zeus ——. Otaheite, S. Parkinson.

NOTES: the fish depicted is a juvenile but the coloration is unusual as the species usually has vertical lines on the sides (Randall, 1955). Chaetodon militaris is probably a Broussonet annotation; this name is one of several in the genus *Chaetodon* which Broussonet communicated to Gmelin (1789). These names were published in a footnote (p.1269) but are not available in nomenclature.

98.(**2**:22*b*) *Chaetodon ulietensis* Cuvier in Cuvier & Valenciennes, 1831 Chaetodontidae

DRAWING: finished water-colour; *r.* [pencil] 'Ch. Falcula/(ulietensis, C.V.)/[ink] S. Parkinson/[pencil] Para harchah nouton [unreadable word]'; *v.* [pencil] 'N°. 67 The fish lost/The snout a little reddish the stripes on the [? front] bright yellow or Orange [unreadable] the tops of the P.D. narrow edge of [unreadable] gray on the

P.A. the [unreadable] or rather grayish [unreadable] on the back [unreadable]/ [ink] Ulhietea'. 233 × 295.

MANUSCRIPT: Solander – (D. & W. 40c) P.A.O.P. not traceable due to absence of a Solander name on the drawing, possibly referred to at f.41 (159) as 'N.B. in Cagg N° 3 two Chaetodon's figured by Mr Parkinson, not described'. Dryander – Catalogue due to the absence of a trivial name on this drawing it is not identifiable.

NOTES: this drawing (or rather a copy of it) was used as the sole source of Cuvier's (1831) name *Chaetodon ulietensis*; it therefore has type status. The drawing was reproduced by Whitehead (1968) as Plate 16, labelled *Chaetodon falcula*. The annotation 'Ch. falcula (ulietensis, C.V.)' is in the hand of Albert Günther; the two names were long treated as synonyms, but Burgess (1978) regards them as distinct. There are no specimens in the British Museum (Natural History) which could be regarded as *Endeavour* material.

99.(**2**:23*a*) *Cheilodactylus* (*Goniistius*) *vestitus* (Castelnau, 1878) Cheilodactylidae

DRAWING: finished pencil, lateral view by H. D. Spöring; *r*. [pencil] 'CHAETODON/gibbosus/Long. natur./[ink] Spőring'; *v*. [ink] 'Endeavours river'. 238 × 328.

MANUSCRIPT: Manuscript: Solander – not traceable in the manuscript P.N.H. (D. & W. 40b). Dryander – Catalogue f.147 as finished without colour, Chaetodon — – N.C. (Nova Cambria), Spőring.

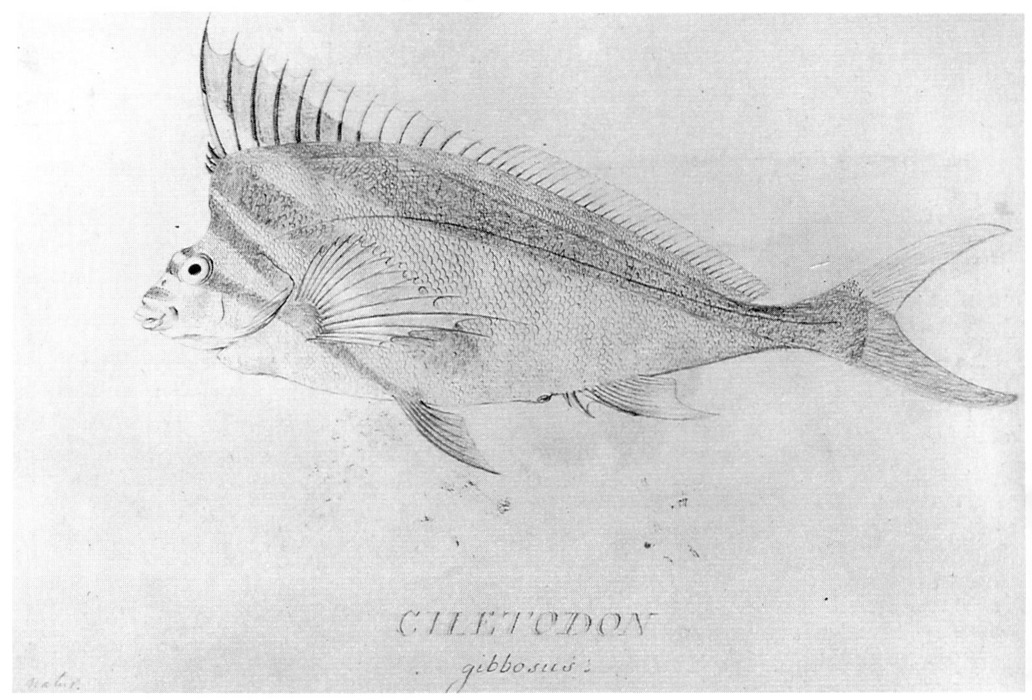

Fig. 8 *Cheilodactylus* (*Goniistius*) *vestitus* (Castelnau, 1878). Drawing by Spöring of a fish from Endeavour River. (Catalogue number 99.)

NOTES: this fish was not described by Solander. Only 15 species of fish were described from New Holland; several of these were drawn by Spöring, a much higher proportion than his zoological drawings elsewhere.

Richardson (1841) described *Cheilodactylus gibbosus* from two specimens collected by John Gould in Western Australia but added '... the fish also inhabits the seas of New Zealand, Mr. Gray having recognised a drawing by Parkinson of a specimen which was caught in Endeavour River, on Cook's second voyage as being a correct representation of this fish. (Vide Banks, Icon. ined. t. 23)'. This suggests that at the date of reading the paper (9 March 1841) Richardson had not himself examined the Parkinson drawings. It also suggests that both Gray and Richardson had confused the first and second Cook voyages, and moreover had misplaced the Endeavour River to New Zealand (rather than present-day Queensland); possibly it was this that caused Richardson to attribute other Australian (New Holland) fishes to the New Zealand fauna (see numbers 49, 50 and 51 in this catalogue). Despite all this confusion it was the apparent identity of the *Endeavour* drawing with Gould's fishes which led Richardson to adopt Solander's manuscript name *Chaetodon gibbosus*, inappropriately as it has recently transpired. Randall (1983) has recently confirmed that *Cheilodactylus gibbosus* Richardson, 1841 is confined to the coast of Western Australia. The morwong which occurs at Endeavour River (and off eastern Australia and in the adjacent Pacific Ocean) is *Cheilodactylus* (*Goniistius*) *vestitus* (Castelnau, 1878).

There are no specimens in the British Museum (Natural History) which can be associated with the *Endeavour* collection.

100. (**2**:23*b*) *Chaetodon trifascialis* Quoy & Gaimard, 1825 — Chaetodontidae

DRAWING: finished water-colour; *r.* [pencil] 'Ch. strigangulus/ [ink] S. Parkinson'; *v.* [pencil] 'N° 6 Chaetodon strigangulus/ [ink] Otahite'. 263 × 326.

MANUSCRIPT: Solander – (D. & W. 40c) P.A.O.P. f.70 (190) as Chaetodon strigangulatus, vernacular names, three specimens preserved, number A128 in Cagg 5, and two specimens A144 in Cagg 6. Dryander – Catalogue f.147, five Parkinson drawings from the Society Islands are listed as sketch with colours Chaetodon; this may be one of the five.

NOTES: the name *Chaetodon striganguli* was published by Gmelin (1789) in a list of species of *Chaetodon* supplied by Broussonet. No doubt Broussonet derived the name from this drawing if not from Solander's manuscript. The name is not available from this date. However, Cuvier (1831) formally published Solander's name, referring to this drawing, and pointing out that the specimen was then in Broussonet's collection. He also referred to Broussonet's communication to Gmelin. Following Cuvier the name *Chaetodon strigangulus* was used by numerous authors (for synonymy see Burgess, 1978).

The Broussonet specimen is still preserved in the Muséum National d'Histoire Naturelle, Paris (MNHN-9680); Bauchot (1969) listed it as *C. trifascialis*. The other specimens from the voyage appear not to have survived; they are not in the British Museum (Natural History).

The drawing was also cited by Lay & Bennett (1839) in an independent proposal of the name *Chaetodon strigangulus*. They also quoted extensively from the Solander manuscript.

This drawing was reproduced in Whitehead (1968) at Plate 18 as *Megaprotodon strigangulus* (Cuvier, 1831).

101.(**2**:24*a*) *Chaetodon citrinellus* Cuvier, in Cuvier & Valenciennes, 1831
Chaetodontidae

DRAWING: finished water-colour; *r.* [pencil] 'Ch. citrinellus mss/Paraharaka eroutoi/[ink] S. Parkinson'; *v.* [pencil] 'to be spotted/where the dotts is, colours sweeten'd in the scales very faint & those on the fins very small – /Nº. 13. Chaetodon punctatus./[ink] Otahite'. 255 × 326.

MANUSCRIPT: Solander – (D. & W. 40c) P.A.O.P. f.15* (132) as Chaetodon punctatus. Dryander – Catalogue f.147 as sketch with colours, Chaetodon citrinellus Brouss —— Soc. Isl., S. Parkinson.

NOTES: the name *citrinellus* was coined by Broussonet for this specimen and it was one of several names in *Chaetodon* which he communicated to Gmelin (1789), although it it not available nomenclaturally from that date. However, Cuvier (1831) validated it citing the Parkinson drawing in Banks's library and Solander's manuscript in addition to other material. He had a specimen from Broussonet, believed to have come from the *Endeavour* collection, which is still preserved in the Muséum National d'Histoire Naturelle, Paris (MNHN-9905) (Bauchot, 1969).

102.(**2**:24*b*) *Chaetodon trifasciatus lunulatus* Quoy & Gaimard, 1824
Chaetodontidae

DRAWING: finished water-colour; *r.* [pencil] 'Ch. vittatus/Seb. th.3 tab.29. f.18. Ch. bellissimus Paraharaka ututhi/[ink] S. Parkinson'; *v.* [pencil] 'Nº. 35[ink] Otahite'. 236 × 296.

MANUSCRIPT: Solander – none. Dryander – Catalogue f.147, possibly the drawing labelled Chaetodon bellus Brouss. —— Society Islands, S. Parkinson.

NOTES: none of the names given in the annotations to the drawing can be identified with entries in Solander's manuscript. The drawing may be the one Dryander listed as *C. bellus* in his *Catalogue*, and this name was one of the several communicated by Broussonet to Gmelin (1789) as *belli*. This drawing was referred to by Cuvier (1831) under his account of *Chaetodon trifasciatus* Park, 1797 as a good illustration by Parkinson. For many years this species has been referred to by Park's name but Burgess (1978) divided the species into two subspecies of which the Pacific form is *C. trifasciatus lunulatus*.

The name *Chaetodon vittatus* published by Bloch & Schneider, 1801, was based solely on Park's description and is thus a junior objective synonym of *C. trifasciatus*.

Lay & Bennett (1839) cited this drawing as *Chaetodon bellus* under their description of *Chaetodon vittatus*, and also drew attention to their inability to find a description in the Solander manuscript.

A specimen in the British Museum (Natural History) (Günther, 1860:24, specimen g) from the Old Collection may be the *Endeavour* specimen. It is 108 mm T.L. (92 mm S.L.).

103.(**2**:25) *Naso unicornis* (Forsskål, 1775) Acanthuridae

DRAWING: finished water-colour; *r.* [pencil] 'Lat- Aumàumā/Eumae/variet Chaet. unicornis ?/ [ink] S. Parkinson'; *v.* [pencil] 'the spots on the P.D. & lines on the P.A. are pale grey – about the middle of the back is a ridge/2 spots of dark blue near the tail. the green brighter on the sides – /N° 5 Chaetodon olivaceous/ [ink] Otahite'. 270 × 371.

MANUSCRIPT: Solander – (D. & W. 40c) P.A.O.P. f.7 (194) as Chaetodon olivaceus, one specimen number A.124, in Cagg 5 (also f.10* (125) another specimen 'unmark'd in ye old Cagg'; the first specimen was the figured one). Dryander – Catalogue f.145, as sketch with colours, C. unicornis var. Brouss. —— Soc. Isl., S. Parkinson.

NOTES: this drawing and Solander's description together served as the sole source of Valenciennes's (1835) name *Naseus olivaceus*. Günther (1861) recognized this species and listed a specimen as 'a. Four inches long. Otaheiti. Type of the species . . .'; this specimen is still preserved in the British Museum (Natural History) at 1962.12.14.3., it is 109 mm T.L. (90 mm S.L.). This illustration was reproduced by Whitehead (1968) as Plate 22; he gives the length of the drawing as 120 mm T.L. Recently remeasured it is 118 mm, which is near enough to the length of the surviving specimen allowing for shrinkage over two hundred years in alcohol, to permit recognition of it as the type specimen.

104.(**2**:26) *Chaetodon* (*Lepidochaetodon*) *unimaculatus unimaculatus* Bloch, 1787 Chaetodontidae

DRAWING: unfinished water-colour; *r.* [pencil] 'Ch. unimaculatus/chaet. fugitivus mss/Paraharahe teare/ [ink] S. Parkinson'; *v.* [pencil] 'N° 18 Chaetodon ocellatus/ [ink] Otahite'. 270 × 373.

MANUSCRIPT: Solander – (D. & W. 40c) P.A.O.P. f.69 (189) as Chaetodon ocellatus, vernacular names, three specimens marked A.126 in Cagg 5, one unmarked (the list of specimens preserved at f.290 gives two specimens marked A.126). Dryander – Catalogue f.147 as sketch with colours, C. fugitivus Brouss. —— Society Islands, S. Parkinson.

NOTES: this drawing and the Solander description were cited by Cuvier (1831) under the description of *Chaetodon unimaculatus* Bloch. Solander's name Chaetodon ocellatus was used independently by Bloch (1781) for a different taxon. The third name in the annotations, Chaetodon fugitivus, was Broussonet's and was published by Gmelin (1789) as *C. fugitivi* although it is not a name available in zoological nomenclature from this date.

Günther (1860) lists a specimen of *Chaetodon unimaculatus* as 'c. Adult: bleached. Old Collection' which may be one of the *Endeavour* specimens, although at 90 mm T.L. (75 mm S.L.) it is too small to be the figured specimen if that was drawn life size. Whitehead (1968) gives the drawing as 116 mm T.L.

105.(**2**:27*a*) *Chaetodon lunula* (Lacepède, 1803) Chaetodontidae

DRAWING: unfinished water-colour; *r.* [pencil] 'Ch. lunula/ [ink] S. Parkinson'; *v.* [pencil] 'Čhaetodon corruscus/ [ink] Princess Island'. 256 × 323.

MANUSCRIPT: Solander – none. Dryander – Catalogue – no drawing listed from Princes Island.

NOTES: a specimen from the Old Collection still exists in the British Museum (Natural History) with the locality Princes Island on its label; it is specimen 'a' of Günther (1860:25). The specimen is in poor condition, is 135 mm S.L., 153 T.L. Adhering under the bottom of the bottle is a label 'Chaetodon corruscus Princes Island'. This seems sufficient evidence to consider it to be an *Endeavour* specimen.

106.(**2**:27*b*) *Centropyge flavisimus* Cuvier in Cuvier & Valenciennes, 1831 Pomacanthidae

DRAWING: unfinished water-colour; *r.* [ink] 'S. Parkinson'; *v.* [pencil] 'N°. 32/ The whole fish is a fine bright orange like the rupicola having a narrow black [border – deleted] at the edge of the/soft part of the P.D. & Pect. & tail. the circle round the eye & the marks by the gills ultramarine./the Iris of the eye brownish gold colour pupil black./Chaetodon luteolus [ink] Otahite'. 235 × 290.

MANUSCRIPT: Solander – (D. & W. 40c) P.A.O.P. f.15* (132) as Chaetodon luteolus. Dryander – Catalogue f.147, presumed to be one of several unspecified drawings of *Chaetodon* by Parkinson from the Society Islands.

NOTES: this drawing was cited by Cuvier (1831) as the sole source of his *Holacanthus luteolus* Cuvier, in Cuvier & Valenciennes, 1831; he appears to have known only the drawing not Solander's manuscript.

107.(**2**:28) *Acanthurus lineatus* (Linnaeus, 1758) Acanthuridae

DRAWING: unfinished water-colour; *r.* [pencil] 'Chaetodon lineatus L./ [ink] S. Parkinson'; *v.* [pencil] 'Amàroa/The edging round the P.D. & Pect. is pale sky blue./N°. 16 Labrus elegantissimus/ [ink] Otahite'. 269 × 378.

MANUSCRIPT: Solander – (D. & W. 40c) P.A.O.P. f.21 (139) as Labrus elegantissimus, specimen A51 (apparently not preserved see f.(287)). Dryander – Catalogue not entered as Labrus elegantissimus, the drawing may have been one of several listed as *Labrus* from the Society Islands, ff.151–153.

NOTES: Solander's manuscript name Labrus elegantissimus seems not to have been used by later workers.

108.(**2**:29*a*) *Zanclus cornutus* (Linnaeus, 1758) Acanthuridae

DRAWING: finished water-colour; *r.* [pencil] 'pale blue [note referring to dorsal fin]/Ch. Cornutus L/Tatàhee/ [ink] S. Parkinson'; *v.* [pencil] 'there is of this fish as large again/N°. 21 Chaetodon rostratus/ [ink] Otahite'. 237 × 294.

MANUSCRIPT: Solander – (D. & W. 40c) P.A.O.P. f.1 (113) as Chaetodon rostratus, several specimens listed at numbers A1, A77 and A162; specimen A77 is noted as 'larger and better'. Dryander – Catalogue f.145 as sketch with colours Chaetodon cornutus L. —— Society Islands, S. Parkinson.

NOTES: the note on the drawing 'there is of this fish as large again' was presumably made with reference to the note in Solander's manuscript that specimen A77 was 'larger and better' than the drawn specimen. A total of four specimens were preserved (Solander, P.A.O.P. f.(285)) on the voyage. Günther (1860:493) listed two specimens of *Z. cornutus* as from the Old Collection (specimens m and n); possibly both were from the *Endeavour* voyage but both have subsequently been deleted from the annotated catalogue and presumably destroyed.

109.(**2**:29*b*) *Chromis chromis* (Linnaeus, 1758) Pomacentridae

DRAWING: finished water-colour; *r*. [ink] 'Chaetodon luridus/ [pencil] castaneus mss/ [ink] S. Parkinson/T.20. Madeira'; *v*. [none]. 145 × 236.

MANUSCRIPT: Solander – (D. & W. 45) S.C. Pisces 1, f.176–177*v*. as Chaetodon luridus, habitat in Oceano Atlantico ad Maderam, vernacular name Castanhete; (D. & W. 42) C.S.D. f.227 (254), same data. Dryander – Catalogue f.147 as finished in colours, Chaetodon luridus Broussonet —— Madeira, S. Parkinson.

NOTES: the name *Chaetodon luridus* was published by Cuvier (1830) and derived from a specimen in Broussonet's collection. This specimen is still preserved in the Muséum National d'Histoire Naturelle (MNHN 5286) S.L. 82 mm (T.L. 110). It is the holotype of *Glyphisodon luridus* Cuvier in Cuvier & Valenciennes, 1830 and is undoubtedly the *Endeavour* specimen from Madeira. The name, as *C. luridi*, was communicated by Broussonet to Gmelin (1789) but is not available for zoological nomenclature as of this date. As Cuvier pointed out, Solander also used the name *Chaetodon luridus* for an acanthurid from Tahiti (see P.A.O.P. f.47 (165), three specimens preserved, A91, A99, and A174), but Cuvier was apparently unaware that the Madeiran fishes were described and listed in the Solander *Slip Catalogue* (D. & W. 45) not in the formal manuscript from the voyage which, it is believed, Cuvier had had copied (see earlier p.23, and Diment & Wheeler (1984)).

110.(**2**:30*a*) *Acanthurus glaucopareius* Cuvier, 1829 Acanthuridae

DRAWING: unfinished water-colour; *r*. [pencil] 'ch. glaucopareius mss./id Seba Thes.3. t25. n.3/Maito/ [ink] S. Parkinson'; *v*. [pencil] 'N°. 62 Chaetodon umbra/ [ink] Otahite'. 258 × 329.

MANUSCRIPT: Solander – (D. & W. 40c) P.A.O.P. f.116, (236) as Chaetodon umbra, one specimen preserved, numbered A179, in Cagg 6. Dryander – Catalogue f.145 as Chaetodon glaucopareius Broussonet, sketch with colours —— Society Islands, S. Parkinson.

NOTES: *Acanthurus glaucopareius* was a name communicated to Cuvier by J. R. Forster, on the basis of a specimen from Cook's *Resolution* voyage which was later named *A. nigricans* Bloch & Schneider, 1801. The *Endeavour* specimen was not referred to in either account.

111.(**2**:30*b*) *Chaetodon vagabundus* Linnaeus, 1758 Chaetodontidae

DRAWING: unfinished water-colour; *r.* [pencil] 'dark chestnut [refers to colouring of caudal fin edge]/Ch. vagabundus/chaet. speciosus mss./Dors. 12 totium/ Parahã/Paraharaha outou rore [indecipherable]/[ink] S. Parkinson'; *v.* [pencil] 'N°. 48 Chaetodon aulicus/[ink] Otahite'. 246 × 341.

MANUSCRIPT: Solander – (D. & W. 40c) P.A.O.P. f.81 (201), vernacular names, one specimen, numbered A136, in Cagg 5. Dryander – Catalogue f.147 as Chaetodon speciosus Broussonet, sketch with colours —— Society Islands, S. Parkinson.

NOTES: both the names C. speciosi and C. aulici were published by Gmelin (1789) in a note communicated by Broussonet, but neither name is available for zoological nomenclature. Cuvier (1831) refers to this drawing, using the name *Chaetodon speciosus*, within the synonymy of *C. vagabundus*, and referring to the citation in Gmelin. The specimen from Broussonet's collection was examined by Cuvier; it is still preserved in the Muséum National d'Histoire Naturelle, Paris (MNHN A 10067) S.L. 89 mm (T.L. 108) (Bauchot, 1969). This is presumed to be the *Endeavour* specimen.

112.(**2**:31*a*) *Heniochus chrysostomus* Cuvier in Cuvier & Valenciennes, 1831 Chaetodontidae

DRAWING: unfinished water-colour; *r.* [pencil] 'chaetodon macrolepidotus L./ Tatoha/Peooè/[ink] S. Parkinson'; *v.* [pencil] 'N°. 54 Chaetodon Chrysostomus/ [ink] Otahite'. 247 × 330.

MANUSCRIPT: Solander – (D. & W. 40c) P.A.O.P. f.64 (184), as Chaetodon chrysostomus, vernacular names, one specimen, numbered A117 in Cagg 5. Dryander – Catalogue f.145, as sketch with colours, Chaetodon macrolepidotus L. —— Society Islands, S. Parkinson.

NOTES: *Heniochus chrysostomus* was described by Cuvier (1831) solely from this drawing by Parkinson and the name was derived from the annotation on it; he made no reference to the Solander manuscript description. This drawing therefore has some type status. Both the drawing and the Solander manuscript were cited by Lay & Bennett (1839) and they quoted extensively from the latter under the name *Heniochus chrysostomus*.

This drawing was reproduced by Whitehead (1968) as plate 17, identified as *Heniochus acuminatus* (Linnaeus, 1758).

113.(**2**:31*b*) *Abudefduf sexfasciatus* (Lacepède, 1801) Pomacentridae

DRAWING: unfinished water-colour; *r.* [ink] 'S. Parkinson/[pencil] Chaet. saxatilis L./Emanmoa'; *v.* [pencil] '73 Chaetodon coelestinus/[ink] Ulhietea'. 249 × 328.

MANUSCRIPT: Solander – (D. & W. 40c) P.A.O.P. f.129, (249) as Chaetodon coelestinus, total of four specimens, one reference number A200 in Cagg 6 (but not marked with the number) and three unnumbered, 'Hab. in Oceano Pacifico prope

Fig. 9 *Kyphosus sectatrix* (Linnaeus, 1766). A Parkinson drawing of a rudder-fish made in the tropical Atlantic in 1768. (Catalogue number 114.)

Insulam Ulhaietea'. Dryander – Catalogue f. 145 as sketch with colours C. saxatilis L. —— Society Islands, S. Parkinson.

NOTES: this drawing and Solander's description together were used by Cuvier (1830) for the name *Glyphisodon coelestinus* Solander, although Cuvier also had other material and discussed other literary references. Cuvier's figure (Plate 135) is reversed from the Parkinson drawing and is slightly smaller but the two are very similar and the plate may have been engraved from a copy of the drawing.

This drawing was reproduced by Whitehead (1968) as Plate 23 *Glyphisodon coelestinus* Cuvier, 1830; as he pointed out in the notes to this plate a specimen listed by Günther (1862) as from the Old Collection, but subsequently destroyed, may have been the *Endeavour* specimen.

114.(**2**:32) *Kyphosus sectatrix* (Linnaeus, 1766) Kyphosidae

DRAWING: finished water-colour; *r.* [ink] 'Chaetodon cyprinaceus/Sydney Parkinson pinx^t ad vivum 1768'; *v.* [ink] 'Novr. 15. 1768 Lat. N.'. 273 × 366.

MANUSCRIPT: Solander – (D. & W. 45) S.C. Pisces 1, f. 166–167v. as Chaetodon cyprinaceus, 'Habitat in Pelago intra Tropicas, ubi Latid. Sept. VI. 50 [i.e. 6° 50′ N.]. Longit. occid. a Londini xxi.7. [i.e. 21° 7′ W of Greenwich] captus Octr. 15. 1768'. (D. & W. 42) C.S.D. f.223 (250) as above. Dryander – Catalogue as Chaetodon cyprinaceous MSS, finished in colours —— Ocean, S. Parkinson.

NOTES: Solander's name *Chaetodon cyprinaceus* was published by Cuvier (1831) in the discussion of the species he called *Pimelepterus boscii* Lacepède. Cuvier cited both the Parkinson drawing and Solander's manuscript description, correctly giving the date as 15 October 1768 (not November as is written on the drawing). The name was one of the several species referred to the genus *Chaetodon* and communicated by Broussonet to Gmelin (1789) who published it as *C. cyprinacei*.

A specimen from Broussonet's collection is still preserved in the Muséum National d'Histoire Naturelle (MNHN-2977), S.L. 121 mm (T.L. 160) originally registered as *Chaetodon cyprinaceus* Broussonet (Bauchot, 1969). Specimens with a similar name were recorded by Günther (1859) under the taxa *Pimelepterus fuscus* ('d. Young. Old Collection, as *Chaetodon cyprinoides*') and *Pimelepterus waigiensis* ('a. Adult. Old Collection, as *Chaetodon cyprinoides*') Although there is no record of more than one specimen being caught on the *Endeavour* voyage the similarity of names suggests they might date from this period.

115.(**2**:33) *Kyphosus incisor* (Cuvier in Cuvier & Valenciennes, 1831) Kyphosidae

DRAWING: finished water-colour; *r.* [ink] 'Chaetodon incisor./Sydney Parkinson pinx^t. 1769'; *v.* [ink] 'Brasil'. 294 × 467.

MANUSCRIPT: Solander – (D. & W. 45) S.C. Pisces 1, f.168–171 as Chaetodon incisor, 'Habitat in Brasilia', length 535 mm ('Diameter longitudinalis 21 unc'). (D. & W. 42) C.S.D. f.225 (252) same data. Dryander – Catalogue f.145 as Chaetodon incisor Mss, finished in colours —— Brasil, S. Parkinson.

NOTES: Cuvier (1831) adopted the Solander name *Chaetodon incisor*, although he attributed it to Parkinson, in his description of *Pimelepterus incisor*, although he had other material in addition. He also cited Solander's manuscript, so this drawing and the associated description have some standing as type material.

This drawing was reproduced by Whitehead (1968) at Plate 14.

116.(**2**:34) *Chaetodipterus faber* (Broussonet, 1782) Ephippidae

DRAWING: finished water-colour; *r.* [ink] 'Chaetodon Gigas./Sydney Parkinson pinx^t 1769'; *v.* [pencil – partly indecipherable notes on coloration in four lines] [ink] 'Brasil'. 288 × 455.

MANUSCRIPT: Solander – (D. & W. 45) S.C. Pisces 1, f.161–163*v* as Chaetodon gigas, 'Habitat in Brasilia at Rio Janeiro', length 20 inches. (D. & W. 42) C.S.D. f.221 (245), same data. Dryander – Catalogue, f.145 as Chaetodon gigas Mss, finished in colours — Brasil, S. Parkinson.

NOTES: although the name *Chaetodon faber* was proposed by Broussonet (1782) he did not associate this drawing or the description with it. His name was based on several earlier literary sources, and specimens (or records of occurrence) from Jamaica and Carolina. He also cited a specimen in 'Mus. Banks' collected in the Society Islands by Banks and Solander; this could not have been the specimen from which the drawing was made which came from Brazil. Broussonet's species is clearly composite as *C. faber* is not found in the Indo-Pacific.

117.(**2**:35) *Abudefduf saxatilis* (Linnaeus, 1758) Pomacentridae

DRAWING: unfinished water-colour; *r.* [pencil] 'Chaetodon/ [ink] S. Parkinson'; *v.* [pencil] 'N° 16 Sparus/ [ink] Brasil'. 271 × 368.

MANUSCRIPT: Solander – (D. & W. 45) S.C. Pisces 1, f.211 as Sparus latus; it is not certain that this entry refers to this drawing, but the note 'Fig. Pict' shows that there was a drawing made of this Brazilian specimen. Dryander – Catalogue f.149 probably the second entry under *Sparus*, sketch with colours, Sparus —— Brasil, S. Parkinson.

NOTES: as there is no trivial name associated with this annotation of *Sparus* it is difficult to relate it to the manuscript. However, for the reasons given above this fish may have been the species Solander named Sparus latus.

118.(**2**:36) *Pseudolabrus miles* (Bloch & Schneider, 1801) Labridae

DRAWING: unfinished water-colour; *r.* [ink] 'S. Parkinson'; *v.* [pencil] '20 Sparus rubecula α/ [ink] Totarranue'. 267 × 371.

MANUSCRIPT: Solander – (D. & W. 40a) P.A., f.6 (8) as Sparus rubecula α, habitat off Cape Kidnappers, one specimen, numbered B9. Dryander – Catalogue f.149 as sketch with colours Sparus —— New Zealand, S. Parkinson.

NOTES: Solander's name was published by Richardson (1843*c*) as *Julis? rubecula*, and he cited Parkinson's drawing deriving the coloration from it and extensively quoted Solander's manuscript description. The drawing therefore has some type standing. Totara nui or Queen Charlotte Sound, South Island, was visited between 15 January and 6 February 1770.

119.(**2**:37*a*) *Pseudolabrus celidotus* (Bloch & Schneider, 1801) Labridae

DRAWING: unfinished pencil sketch; *r.* [ink] 'S. Parkinson'; *v.* [pencil] '16 Sparus notatus/ [ink] Totarra nue'. 257 × 329.

MANUSCRIPT: Solander – (D. & W. 40a) P.A. f.12 (14), as Sparus notatus habitat Tolaga, five specimens preserved with the serial number B.17. Dryander – Catalogue f.149 as sketch without colours Sparus —— New Zealand, S. Parkinson.

NOTES: Richardson (1843*c*) adopted Solander's name for his *Julis? notatus* and extensively quoted his description, as well as referring to this drawing. As he pointed out Parkinson's drawing is labelled Totarra nue (in Queen Charlotte's Sound) while Solander's description is of a fish from Tolaga (38°20′S, 178°21′E) on North Island, where the *Endeavour* was between 23 and 29 October 1769. This drawing therefore has some type status.

It was reproduced by Whitehead (1968) as Plate 25A.

120.(**2**:37*b*) *Diplodus sargus* (Linnaeus, 1758) Sparidae

DRAWING: finished water-colour by A. Buchan; *r.* [ink] 'Sparus Sargus/Buchan'; *v.* [ink] 'Madera'. 185 × 271.

MANUSCRIPT: Solander – (D. & W. 45) S.C. Pisces 1, f. 193–194v as Sparus sargus, habitat in Mari Atlantico Madera, length 4 inches. (D. & W. 41) F.C. f. 229 (256), same data. Dryander – Catalogue f. 149 as Sparus sargus, finished in colours —— Madeira, Buchan.

NOTES: this well-known European sea bream was briefly described by Solander; he included the Madeiran vernacular name 'Sargo' interpreting it parenthetically as 'accute or cunning fellow'.

121. (**2**:38) *Pseudolabrus miles* (Bloch & Schneider, 1801) Labridae

DRAWING: unfinished water-colour; *r.* [ink] 'S. Parkinson'; *v.* [pencil] '4 Sparus rubiginosus β/ [ink] Mattaruwhow'. 272 × 331.

MANUSCRIPT: Solander – (D. & W. 40a) P.A. f. 7 (9) as Sparus rubiginosus var β, habitat off Cape Kidnappers, one specimen labelled B.8. (Also references to S. rubiginosus var α at P.A., f. 7 (9) and P.A.O.P. f. 47 (123) (D. & W. 40c).) Dryander – Catalogue f. 149, probably one of the sketches without colour (erroneously) Sparus —— New Zealand, S. Parkinson.

NOTES: Richardson (1843*c*) published Solander's name as *Julis? rubiginosus*, quoting his description extensively and referring to Parkinson's drawing. The drawing therefore has some type status. As Richardson pointed out Parkinson's figure was drawn from a specimen taken at Mattaruhow and Solander's description from a specimen off Cape Kidnappers.

This drawing was reproduced by Whitehead (1968) as Plate 25B.

122. (**2**:39) *Archosargus rhomboidalis* (Linnaeus, 1758) Sparidae

DRAWING: unfinished water-colour by A. Buchan; *r.* [pencil] 'the orange more of a Golden Colour/the white Like silver mixt with purple/ [ink] Buchan'; *v.* [pencil] 'N°. 7. Sparus rhomboides/ [ink] Brasil'. 268 × 363.

MANUSCRIPT: Solander – not found. Dryander – Catalogue f. 149 as sketch with colours Sparus —— Brasil, Buchan.

NOTES: the annotations correcting the colouring on the drawing help confirm the identification of this fish.

123. (**2**:40) *Nemadactylus macropterus* (Bloch & Schneider, 1801) Cheilodactylidae

DRAWING: unfinished water-colour; *r.* [pencil] 'Sparus carponemus mss/ [ink] S. Parkinson'; *v.* [pencil] '3. Sciaenoides abdominalis [ink] Queen Charlottes Sound'. 293 × 477.

MANUSCRIPT: Solander – (D. & W. 40a) P.A. f. 29 (31) and f. 9 (11), as Sciaena abdominalis, both with 'Fig. Pict.' hence suggesting there were two illustrations (see no. 138). Dryander – Catalogue f. 155 as sketch with colours, Sciaena [in pencil Sparus carponemus Brouss.] —— New Zealand, S. Parkinson.

NOTES: this species was drawn and described twice. Firstly, off 'Cape Kidnappers' (f.9 (11)) where two specimens were preserved numbered B.5. Secondly, in Motuaro Bay on 24 November 1769, stated Longitude LXXX W. (f.29 (31)). The second drawing is no. 138 of this catalogue.

This drawing was referred to by Cuvier (1830) under his account of *Cheilodactylus carponemus* as his reference to Queen Charlotte Sound proves. However, this locality is spurious, as discussed under number 138 of this catalogue.

124.(**2**:41) *Anamses coeruleopunctatus* Rüppell, 1829 Labridae

DRAWING: unfinished water-colour; *r.* [pencil] 'Padee/[ink] S. Parkinson'; *v.* [pencil] 'The whole colour of the fish is darker blue done w^t^ Ultramarine/the red there a cast of Green especialy on the back towards the tail./N°. 24 Sparoides azureus/[ink] Otahite'. 269 × 371.

MANUSCRIPT: Solander – (D. & W. 40c) P.A.O.P. f.50 (170) as Sparoides azureus, one specimen numbered A.98 in Cagg 4. Dryander – Catalogue f.149 [pencil entry], sketch with colours, Sparoides —— Society Island, S. Parkinson.

NOTES: Solander's name *Sparoides azureus* does not seem to have been taken up by any later author. Two specimens in the British Museum collection listed by Günther (1862) as 'c, d. Adult: not good state.' might possibly have been *Endeavour* material but have been destroyed since Günther wrote.

125.(**2**:42) *Calotomus carolinus* (Valenciennes in Cuvier & Valenciennes, 1840) Scaridae

DRAWING: unfinished water-colour; *r.* [pencil] 'Êuhoo uelha/[ink] S. Parkinson'; *v.* [pencil] 'the head blue green streaked w^t^ red the body redish brown inclining to purple towards/the base of each scale red darker on the back & lighter on the belly./ N°. 20 Callyodon pictus'. 268 × 373.

MANUSCRIPT: Solander – (D. & W. 40c) P.A.O.P. f.15 (133) as Callyodon pictus, one specimen numbered 37. Dryander – Catalogue f.151, sketch with colours, Callyodon —— Society Islands, S. Parkinson.

NOTES: the Solander name Callyodon pictus seems not to have been taken up by subsequent authors.

126.(**2**:43) *Euscarus cretensis* (Linnaeus, 1758) Scaridae

DRAWING: finished water-colour; *r.* [ink] 'Callyodon rubiginosus/Sydney Parkinson pinx^t^ 1768'; *v.* [ink] 'Madera'. 237 × 294.

MANUSCRIPT: Solander – (D. & W. 45) S.C. Pisces 1 f.216–218v as Callyodon rubiginosum, 'Habitat in Oceano Atlantico prope Insulam Maderam', specimen 5½ inches long. (D. & W. 42) C.S.D. f.239 (266), same data. Dryander – Catalogue f.151 as Callyodon rubiginosus MSS, finished in colours —— Madeira, S. Parkinson.

NOTES: the name *Scarus rubiginosus* was adopted from Solander's manuscript by Valenciennes (1840). Valenciennes cited the vernacular name 'budiam' from Solander's manuscript, or a copy of it, and also referred to the Parkinson drawing.

Other, published, accounts were cited by Valenciennes, but this drawing can be considered to have some standing as type material. The eastern tropical Atlantic parrotfish occurring at Madeira is usually considered to be conspecific with the eastern Mediterranean *E. cretensis*.

127.(**2**:44) *Coridodax pullus* (Bloch & Schneider, 1801) Odacidae

DRAWING: unfinished water-colour; *r.* [pencil] 'Callyodon coregonoides/[ink] S. Parkinson'; *v.* [pencil] 'the Strip on the side silvery the spots on the P.D. & P.A. transparent the membranes of/the tail transparent the spots on the side a purple gray./2 Coregonoides vittatus/[ink] Mattaruwhow'. 294 × 479.

MANUSCRIPT: Solander – (D. & W. 40a) P.A., f.3 and 52 (54) as Coregonoides vittatus. Dryander – Catalogue, not found.

NOTES: there are two entries in the Solander manuscript referring to this species, f.3 refers to two specimens with the serial number B1, and Fig. Pict. at this entry suggests that one of these was the illustrated specimen, while the entry at f.52 appears to have no precise locality and there is no evidence that specimens were preserved on this later occasion.

Solander's name vittatus was published by Richardson (1843*c*) as *Odax vittatus* with an extensive quotation of Solander's manuscript description and reference to Parkinson's drawing. This name is based solely on this source.

The reference by Valenciennes (1839) to *Labrus cyanogaster* Solander from Tahiti does not appear to refer to this species although Whitehead (1968) suggested that it might.

Whitehead (1968) reproduced this drawing at Plate 26, identifying it as *Thalassoma purpurea* (Forsskål, 1775).

128.(**2**:45) *Scarus psittacus* Forsskål, 1775 Scaridae

DRAWING: unfinished water-colour; *r.* [pencil] 'Callyodon/Paguhoo/Toou epatee/[ink] S. Parkinson'; *v.* [pencil] 'The colour on the back turns paler as it goes towards the Belly./N° 10 Labrus ornatus/[ink] Otahite'. 269 × 369.

MANUSCRIPT: Solander – (D. & W. 40c) P.A.O.P. f.54 (174) as Labrus ornatus, one specimen preserved numbered A107 in Cagg 4. Dryander – Catalogue f.151 as Calliodon, sketch with colours —— Society Islands, S. Parkinson.

NOTES: the name Labrus ornatus does not seem to have been employed by later naturalists.

129.(**2**:46) *Upeneichthys porosus* (Cuvier in Cuvier & Valenciennes, 1829) Mullidae

DRAWING: unfinished water-colour; *r.* [ink] 'S. Parkinson' [in pencil on the mount – 'Upeneus vlamingi Cuv. & Val. t.3. p.453']; *v.* [pencil] 'the part mark'd 2 on the face [the preorbital region] is pale green/The belly pale crimson spotted all over with yellow the spots on the base of the scale somewhat/deeper. the streaks marked x [also on the preorbital region] so on the face, the spotts on the back & on the PD. &

Fig. 10 *Upeneichthys porosus* (Cuvier *in* Cuvier & Valenciennes, 1829). An unfinished Parkinson drawing made off Motuaro, New Zealand. The verso of this sheet bears extensive notes on coloration for the completion of the drawing. (Catalogue number 129.)

PA the/outer circle of the eye & streaks on the tail, ultramarine w[t] a cast of Purple, the streaks on the face & spots on the back being the deepest./11 Labrus calopthalmus/ [ink] off Motuaro'. 269 × 367.

MANUSCRIPT: Solander – (D. & W. 40a) P.A. f.40 (42), as Labrus calophthalmus, habitat off Motuaro, one specimen preserved numbered B23. Dryander – Catalogue f.153, as Labrus —— sketch with colours, New Zealand, S. Parkinson, the last entry under *Labrus*.

NOTES: this drawing was cited by Cuvier (1829) under the name *Upeneus vlamingii* and he gave a short description of the species based on it. Cuvier reports the locality as Queen Charlotte's Sound.

130.(**2**:47*a*) *Thalassoma pavo* (Linnaeus, 1758) Labridae

DRAWING: finished water-colour; *r.* [ink] 'Labrus lunarius/Sydney Parkinson pinx[t]. 1768.'; *v.* [ink] 'Madera'. 236 × 295.

MANUSCRIPT: Solander – (D. & W. 45) S.C. Pisces 2, f.8–11 as Labrus lunaris, Habitat in Oceano Atlantico Insulam Maderam alluente, length 6 inches. (D. & W. 42) C.S.D. f.243 (270), same data. Dryander – Catalogue f.151 as *Labrus lunaris* L., finished in colours, —— Madeira, S. Parkinson.

NOTES: this common Mediterranean and tropical eastern Atlantic Ocean wrasse was confused by Solander with *Labrus lunaris* Linnaeus, 1758, an Indo-Pacific species. His description of the specimen was very detailed.

131.(**2**:47*b*) *Thalassoma lutescens* (Lay & Bennett, 1839) Labridae

DRAWING: unfinished water-colour; *r.* [pencil] 'Labrus lorius mss./Epaou pararoute/[ink] S. Parkinson'; *v.* [pencil] 'N°. 58. Labrus lutescens/[ink] Otahite'. 238 × 295.

MANUSCRIPT: Solander – (D. & W. 40c) P.A.O.P. f.49 (169) as Labrus lutescens, one specimen numbered A96 in Cagg 4. Dryander – Catalogue f.151 as sketch with colour, Labrus lorius Brouss. —— Society Islands, S. Parkinson.

NOTES: the Solander name was employed by Lay & Bennett (1839) in their description of *Julis lutescens* quoting extensively from Solander's manuscript account. With permission (presumably of Robert Brown although this was not stated) they reproduced the Parkinson drawing as their Plate XIX, Figure 2; the drawing being finished by the completion of fin rays in the fins and scales on the body (the original has only samples of both shown). They did not, however, add the pelvic fins which were omitted in Parkinson's drawing although mentioned by Solander. Lay and Bennett's text was prepared at least nine years before publication. Therefore the Parkinson drawing must have been copied around 1830 (Beechey, 1839).

This drawing was reproduced by Whitehead (1968) as Plate 27, labelled *Thalassoma lunaris* (Linnaeus, 1758).

132.(**2**:48) *Thalassoma quinquevittata* (Lay & Bennett, 1839) Labridae

DRAWING: unfinished water-colour; *r.* [pencil] 'Paöou-marouruh/páou-móurúrá/[ink] S. Parkinson'; *v.* [pencil] 'N°. 8 [ink] Labrus formosus/Otahite'. 270 × 369.

MANUSCRIPT: Solander – (D. & W. 40c) P.A.O.P. f.108 (228), as Labrus formosus, two specimens preserved, numbered A169 and A189, both in Cagg 6. Dryander – Catalogue f.152–3 not identifiable among the ten Parkinson drawings of *Labrus* made at the Society Islands.

NOTES: this drawing was cited by G. T. Lay & E. T. Bennett (1839) as *Labrus formosus*, Sol. MSS in their description of *Scarus quinque-vittatus*, which was based, however, solely on a drawing made by Beechey during the voyage of the *Blossom*. Their account was written before 1830 (Beechey, 1839). Valenciennes (1839) also referred to this drawing, but wrongly attributed it to Forster, as Günther (1862) has already pointed out, in his account of *Julis erythrogaster*; a name which occurs in the Solander manuscripts as *Labrus vittatus erythrogaster* f.7 (119). Günther regarded this, *Labrus vittatus cyanogaster*, and *L. formosus* as varieties of *Julis trilobata* (Lacepède, 1802) and clearly consulted the drawings and manuscripts to reach this decision. There were no specimens extant in the British Museum collection which could have been associated with the *Endeavour* collection (Günther, 1862).

133.(**2**:49) *Thalassoma hardwickei* (Bennett, 1829) Labridae

DRAWING: unfinished water-colour; *r.* [pencil] 'Paäehoe/Epàoū maraoùrā [ink] S. Parkinson'; *v.* [pencil] 'The eyes gold colour & the pupil black – the P.P. transparent the [spots – deleted] stripes on the back soften'd in. – /N°. 30. Labrus pulcherrimus/ [ink] Otahite'. 237 × 295.

MANUSCRIPT: Solander – (D. & W. 40c) P.A.O.P. f.6*, (121) as Labrus pulcherrimus, apparently four specimens preserved numbered A.18, A.119, and A.153 (two fishes) in Caggs 1, 5, and 6 respectively. Dryander – Catalogue f.152–3 not identifiable among the ten Parkinson drawings of *Labrus* made at the Society Islands.

NOTES: Solander's name Labrus pulcherrimus seems not to have been taken up by any later author.

134.(**2**:50*a*) *Paracirrhites forsteri* (Bloch & Schneider, 1801) Cirrhitidae

DRAWING: unfinished water-colour; *r.* [ink] Cirrhites forsteri/ [pencil] perca cruenta mss Seb. thes. 3 tab. 27 n. 12./Taiboo/Ideeiaio/ [ink] S. Parkinson'; *v.* [pencil] 'N°. 42 Labrus rufus/ [ink] Otahite'. 250 × 331.

MANUSCRIPT: Solander – (D. & W. 40c) P.A.O.P. f.53 (173) as Labrus rufus, one specimen preserved, numbered A.97 in Cagg 4. Dryander – Catalogue f.153, as Labrus, in pencil Perca cruenta Brouss. —— Society Islands, S. Parkinson, sketch with colours.

NOTES: the name Labrus rufus Solander seems not to have been adopted by any later naturalist possibly because it was a homonym of *Labrus rufus* Linnaeus, 1758. However, Broussonet's manuscript name Perca cruenta has received considerable usage and resulted in confusing the history of the earliest collected hawk-fishes in that all of them from both first and second Cook voyages were renamed alike. Specimens in the British Museum from the Old Collection were listed by Günther (1860) as *Perca cruentata* and Wheeler (1981) showed that one of these is probably Forster's *Resolution* specimen and thus the type specimen of this species. The second of Günther's specimens was referred to *Cirrhites arcatus* at a later date. A specimen of *Paracirrhites* in the Muséum National d'Histoire Naturelle, Paris (MNHN–A.2912) S.L. 102 mm (T.L. 121), was accessioned as *Cirrhites maculatus* = *Perca cruenta* Broussonet (Bauchot, 1969) and may possibly be the *Endeavour* specimen.

135.(**2**:50*b*) *Halichoeres radiatus* (Linnaeus, 1758) Labridae

DRAWING: finished water-colour by A. Buchan; *r.* [ink] 'Buchan'; *v.* [pencil] 'N°. 14 Labrus/ [ink] Brasil'. 241 × 326.

MANUSCRIPT: Solander – not found. Dryander – Catalogue f.151 as finished in colours, Labrus —— Brasil, Buchan.

NOTES: as this fish appears not to have been described or named by Solander the drawing has not been referred to by later naturalists.

136.(**2**:51*a*) *Stethojulis bandanensis* (Bleeker, 1851) Labridae

DRAWING: unfinished water-colour; *r.* [pencil] 'Labrus Taeniatus mscr./cfr typ. specimen,/perh. new species/Pauuhe/[ink] S. Parkinson'; *v.* [pencil] 'N°. 53 Labrus aulicus/[ink] Otahite'. 247 × 333.

MANUSCRIPT: Solander – (D. & W. 40c) P.A.O.P. f.62 (182), as Labrus aulicus, vernacular names, apparently three specimens numbered A.114, A.154, and A.115 preserved in Caggs 5 and 6 (the two last); a fourth entry to A.115 is deleted possibly because it suggested the specimen was in Cagg 5 not 6. Another reference to Labrus aulicus occurs on f.90 (210) of P.A.O.P. but is deleted and vittatus erythrogaster substituted. Labrus taeniatus occurs on f.51 (171) of P.A.O.P. as a Solander name, but the name on the drawing probably refers to an independent usage by Broussonet. Dryander – Catalogue f.151 as sketch with colours, Labrus taeniatus Brouss. —— Society Islands, S. Parkinson.

NOTES: Solander's name Labrus aulicus seems not to have been adopted by later naturalists, nor apparently was Labrus taeniatus. This latter name was proposed by Broussonet (see Dryander Catalogue entry, above) and presumably he wrote it on the drawing. It too was never published.

137.(**2**:51*b*) *Pagellus bogaraveo* (Brunnich, 1768) Sparidae

DRAWING: finished ink drawing by A. Buchan; *r.* [pencil] 'Positio squamarum non recte./[ink] Labrus [deleted] Sparus [inserted in pencil] griseus/[pencil] Sp. Bogaraveo Brunn. ich mass./p.49/[ink] Buchan'; *v.* [ink] 'Madeira'. 186 × 271.

MANUSCRIPT: Solander – (D. & W. 45) S.C. Pisces 1, f.200–203 as Sparus griseus, habitat in Oceano Maderae, length 4 inches; (D. & W. 42) C.S.D. f.233 (260), same data. Dryander – Catalogue f.149 Sparus – no entry for this drawing.

NOTES: the name Sparus griseus Solander was never taken up by later naturalists, and it seems that at an early date (the annotation may be Broussonet's) this drawing and Solander's description may have been recognized as a juvenile of the red sea-bream, *Pagellus bogaraveo*.

This drawing was reproduced by Whitehead (1968) as Plate 30B.

138.(**2**:52) *Nemadactylus macropterus* (Bloch & Schneider, 1801) Cheilodactylidae

DRAWING: unfinished water-colour; *r.* [pencil] 'N.B./Sparus Carponemus mss/[ink] S. Parkinson'; *v.* [pencil] '3 Sciaenoides abdominalis/[ink] Mattaruwhow'. 270 × 371.

MANUSCRIPT: Solander – (D. & W. 40a) P.A. f.29 (31) and f.9 (11), as Sciaena abdominalis both with Fig. Pict. showing that the species was drawn twice (see number 123, this catalogue). Dryander – Catalogue f.155 presumably the second sketch with colours from New Zealand by S. Parkinson under the heading (four lines above) in pencil Sparus carponemeus Brouss.

NOTES: Sciaena abdominalis was figured and described twice in Solander's Pisces Australiae. At f.9 (11) he recorded the locality as off Cape Kidnappers, a name used in addition to Mattaruwhow (near the present Napier, East coast of North Island). At f.29 (31) he recorded the locality as Motuaro Bay and the date as 24 November 1769 (thus in Huraki Gulf, towards the north of North Island). This drawing therefore appears to have been the first of the two produced. It was reproduced by Whitehead (1968) as Plate 20.

The other drawing (number 123 in this catalogue) is labelled Queen Charlottes Sound (north end of South Island) but if the Solander manuscript localities are correct (and there is no reason to doubt this) then Queen Charlottes Sound is an incorrect locality. This suggestion is more probable than there being an error in Solander's manuscripts as these localities were added to the drawings by Dryander long after the voyage.

Cuvier (1830) used the name *Cheilodactylus carponemus*, deriving it from the Parkinson drawing, which he cited. As he quoted Queen Charlotte Sound as a locality there is no doubt that he was referring to the other drawing (number 123) which has this incorrect locality. Whitehead (1968) was misled by the confusion between the names and localities and claimed that the present drawing was part of the type materials for *C. carponemus*. The point is not important as *C. carponemus* was based largely on the drawings of G. A. Forster (one of which was made at Queen Charlotte's Sound) and J. R. Forster's description which was used by Schneider for the sole basis of his *Cichla macroptera* (see Wheeler, 1981). Cuvier's name is thus best regarded as a junior objective synonym of *Nemadactylus macropterus* (Bloch & Schneider, 1801). This drawing is discussed and Solander's manuscript quoted extensively by Richardson (1842*b*).

139.(**2**:53) ?*Diplodus caudimaculata* (Poey, 1861) Sparidae

DRAWING: unfinished water-colour; *r.* [pencil] '(Sparus latus/plumbeus [written above this] Catalogue)/[ink] S. Parkinson/[pencil] Labrus plumbeus/sent from Jamaica by Shakespear/11. New Zealand Sparus vel Sciena'; *v.* [pencil] 'N° 13. Labrus/[ink] Brasil'. 295 × 457.

MANUSCRIPT: Solander – not found. Dryander – Catalogue f.151 as sketch with colours, Labrus —— Brasil, S. Parkinson, probably refers to this drawing.

NOTES: the absence of a trivial name on this drawing makes it impossible to check in the Solander manuscripts; indeed, it is probable that Solander did not describe this fish (other Brazilian specimens were drawn but not described). The annotations on the drawing are confusing. I believe that those on the verso are the original notes, except that 'Brasil' was added by Dryander. It is possible that the note 'Sparus latus, plumbeus Catalogue' is in the hand of Broussonet. I do not recognize the small, neat hand referring to '11. New Zealand Sparus vel Sciena', nor 'Labrus plumbeus sent from Jamaica by Shakespear'. Roger Shakespear was employed by Banks as a collector of plants; he also sent many specimens of Jamaican fishes to the British Museum where they were examined and described by Solander (Dawson, 1958; Wheeler, 1984).

I interpret the later annotations as follows. The note labelling this Sparus latus possibly by Broussonet was an error in attempting to identify this fish with Solander's ms Sciena lata (see no. 161, this catalogue), but this led to the locality of New Zealand from a later note writer. The reference to Shakespear's Jamaican Labrus plumbeus can be seen merely as a comparison with a specimen received from that collector either when it came to England or already in the British Museum.

140. (**2**:54) *Parapercis colias* (Bloch & Schneider, 1801) Mugiloididae

DRAWING: unfinished water-colour; *r.* [pencil] 'Perca ? colias/ [ink] S. Parkinson'; *v.* [pencil] '7 Labrus macrocephalus/ [ink] Motuaro'. 293 × 479.

MANUSCRIPT: Solander – (D. & W. 40a) P.A. f. 26 (28), as Labrus macrocephalus, 'cole fish nostratibus'. Dryander – Catalogue f. 153, as sketch with colours, Labrus —— New Zealand, S. Parkinson.

NOTES: the Solander name Labrus macrocephalus does not seem to have been taken up by later authors. This species was named from the description made by J. R. Forster on the *Resolution* voyage; Cuvier (1829) referred to the drawing by G. A. Forster of this specimen in Banks's collection but made no reference to the Parkinson drawing from the *Endeavour* voyage. This drawing was referred to by Richardson (1843*a*, 1843*b*); in the latter reference he quoted from Solander's manuscript and compared the description and figures from both the *Endeavour* and *Resolution* voyages.

141. (**2**:55) *Cheilinus trilobatus* (Lacepède, 1802) Labridae

DRAWING: unfinished water-colour; *r.* [pencil] 'Epupoi/ [ink] S. Parkinson'; *v.* [pencil] 'The finns & tail are all strew'd over wt small white spots/every scale is tip'd more or less wt the pink Colour – the darker/red spots are dropt here & there throughout the whole body/74 Labrus cruentus/ [ink] Ulhietea'. 273 × 373.

MANUSCRIPT: Solander – (D. & W. 40c) P.A.O.P. f. 131 (251) as Labrus cruentatus, one specimen preserved, numbered A. 201, 'Habitat in Oceano pacifico prope Insulam Ulhaietea'. Dryander – Catalogue f. 152–3, not distinguishable among the numerous entries Labrus —— Society Island, S. Parkinson.

NOTES: the Solander name does not seem to have been used by later workers.

142. (**2**:56) *Cheilinus trilobatus* (Lacepède, 1802) Labridae

DRAWING: unfinished water-colour; *r.* [pencil] 'Marrarra/ [ink] S. Parkinson'; *v.* [pencil] 'the green in the P.D. & P.A. turns very dark toward the bottom./68/ [ink] Ulhietea'. 270 × 371.

MANUSCRIPT: Solander – not found. Dryander – Catalogue not found.

NOTES: as there is no manuscript name on this drawing it is not possible to identify it in Solander's manuscript. However, there were only five species of fish recorded from Ulhaietea (= Raiatea) including the Labrus cruentatus noted above (no. 141, this catalogue). It may be that the two fish were caught close together in time and both were drawn but only one described. The manuscript list of specimens (P.A.O.P. f.292) records only the single specimen preserved as noted above. It is not now in the collection of the British Museum (Natural History).

143.(**2**:57) *Thalassoma purpureum* (Forsskål, 1775) Labridae

DRAWING: unfinished water-colour; *r.* [pencil] 'Pao-Manulka/[ink] S. Parkinson/[in ink on mount] see Julis quadricolor Lesson Voy Coq:t'; *v.* [pencil] 'The green a deal brighter as if done wt Verdigrease & yellow upon the back two or three fascia of/green clear. the colours must be sweetten'd in dark purple brown on the back the edges of the/scales here & there green./This drawing alter'd the red brighter where the black lead mark is on the/head a fine rose colour the greens more blue./N° 17 Labrus vitatus/Cynogaster [written above vittatus] [ink] Otahite'. 269 × 371.

MANUSCRIPT: Solander – (D. & W. 40c) P.A.O.P. f.12 (128), as Labrus vittatus cyanogaster from George Land (= Tahiti), it is uncertain how many specimens were preserved, 3 according to the index f.(285). Dryander – Catalogue f.151–3 not recognizable among the numerous Labrus —— drawn by Parkinson in the Society Islands.

NOTES: Günther (1862) referred to the Parkinson drawings and the Solander manuscript and corrected Valenciennes (1839) who had wrongly attributed this and two other drawings to G. Forster. However, Günther considered that Solander had had three samples of what he called *Julis trilobata* (Lacepède, 1802) which had been unnecessarily recognized as three taxonomic forms, *Labrus formosus* (see no. 132, this catalogue), *Labrus vittatus cyanogaster* (this drawing), and *Labrus vittatus erythrogaster*, originally Labrus aulicus (no. 144, this catalogue). The concept of subspecies which Günther attributed to Solander in this way was unknown to Solander and the trinomials are a misreading and misunderstanding of Solander's attempt to change the name of *Labrus vittatus* to *Labrus cyanogaster*, and to *L. erythrogaster* for no. 136.

This drawing served as the basis for *Julis cyanogaster* Valenciennes, in Cuvier & Valenciennes, 1839, a name which he attributed to Solander even though he assumed the painting was by Forster.

144.(**2**:58) *Thalassoma fuscum* (Lacepède, 1802) Labridae

DRAWING: unfinished water-colour; *r.* [pencil] 'Epóu páá/[ink] S. Parkinson'; *v.* [pencil] 'Labrus vitatus/The iris yellowish green the pupil black/N° 9 Labrus [aulicus – deleted] vittatus erythrogus [added]/[ink] Otahite'. 268 × 371.

MANUSCRIPT: Solander – (D. & W. 40c) P.A.O.P. f.90 (210) as Labrus aulicus, which deleted and vittatus/erythrogaster substituted; one specimen numbered

A.149, in Cagg number 5. Dryander – Catalogue f.151–3, presumed to be one of several drawings of Labrus —— by Parkinson from the Society Islands.

NOTES: this drawing and the Solander manuscript were examined by Günther (1862). No later naturalist appears to have used Solander's names.

145.(**2**:59) *Halichoeres trimaculatus* (Quoy & Gaimard, 1834) Labridae

DRAWING: unfinished water-colour; *r.* [pencil] 'Epou taa taa/ [ink] S. Parkinson'; *v.* [pencil] 'N°. 25 Labrus osmeroides/ [ink] Otahite'. 272 × 370.

MANUSCRIPT: Solander – (D. & W. 40c) P.A.O.P. f.5 (117) as Labrus osmeroides, locality George Land (= Tahiti), two specimens numbered A.8 and A.108. Dryander – Catalogue f.151–3, presumed to be one of several drawings of Labrus —— by Parkinson from Society Islands.

NOTES: Solander's name Labrus osmeroides seems not to have been used by later naturalists.

146.(**2**:60) *Variola louti* (Forsskål, 1775) Serranidae

DRAWING: unfinished water-colour; *r.* [pencil] 'P. Rosea mss. Perca Louti forsk. fn. arab. p.40. n.40/Ehowau'; *v.* [pencil] 'N°. 1 Perca rosea/ [ink] Otahite'. 298 × 476.

MANUSCRIPT: Solander – (D. & W. 40c) P.A.O.P. f.35 (153) as Perca rosea, one specimen numbered A.69 in Cagg 3. Dryander – Catalogue f.157 as sketch with colours, Perca rosa Brouss —— Society Islands, S. Parkinson.

NOTES: this drawing is the sole source for Cuvier's (1828) *Serranus roseus* which is briefly described from a Parkinson drawing from Tahiti. As an exception to his usual practice Cuvier does not credit the name to Solander, nor does he refer to Solander's manuscript.

There are no specimens in the British Museum (Natural History) which can be associated with the *Endeavour* voyage.

147.(**2**:61) *Epinephelus fasciatus* (Forsskål, 1775) Serranidae

DRAWING: unfinished water-colour; *r.* [ink] 'Serranus fasciatus, Forsk./ [pencil] Whapoo/Matapoo-Ohŏå/ [ink] S. Parkinson'; *v.* [pencil] 'N°. 59. Perca rubescens/ [ink] Otahite'. 269 × 371.

MANUSCRIPT: Solander – (D. & W. 40c) (P.A.O.P. f.123 (243) as Perca rubescens, two specimens, one unmarked, the other numbered A.186 in Cagg 6. Dryander – Catalogue f.157–9, presumed to be one of seven unfinished water-colours of *Perca* —— by Parkinson in the Society Islands.

NOTES: the Solander name Perca rubescens does not appear to have been taken up by later naturalists.

A specimen of *Epinephelus fasciatus* in the Muséum National d'Histoire Naturelle, Paris, which originated in Banks's collection and was given to Broussonet is believed to be a specimen collected on the *Resolution* voyage by J. R. Forster (Bauchot, 1969; Wheeler, 1981).

148.(**2**:62) *Monotaxis grandoculis* (Forsskål, 1775) Pentapodidae

DRAWING: unfinished water-colour; *r.* [pencil] 'Spharodon grandoculis/an Perca Gobioides mss/Emoco/[ink] S. Parkinson'; *v.* [pencil] 'N°. 12 Sciaena cyprinacea [ink] Otahite'. 273 × 370.

MANUSCRIPT: Solander – (D. & W. 40c) P.A.O.P. f.67 (187) as Sciaena cyprinacea, vernacular name, length 16 inches. Dryander – Catalogue f.155, presumed to be one of seven drawings by Parkinson from the Society Islands labelled Sciaena.

NOTES: Solander's index to the P.A.O.P. (f.290) shows that no specimen of Sciaena cyprinacea was preserved. Solander's manuscript name does not appear to have been taken up by any later naturalist. Perca gobioides does not appear in the Solander manuscript and was probably a Broussonet name.

149.(**2**:63*a*) *Holocentrus ascensionis* (Osbeck, 1765) Holocentridae

DRAWING: finished water-colour; *r.* [ink] 'Sciaena rubens./Sydney Parkinson pinx[t] 1768'; *v.* [ink] 'Nov[r]. 8. 1768/Coast of Brasil'. 258 × 324.

MANUSCRIPT: Solander – (D. & W. 45) S.C. Pisces 2, f.34–37 as Sciaena rubens, 'Habitat . . . a territoria Spiritus Sancti Brasiliae'; specimen 8½ inches long. (D. & W. 42) C.S.D. f.249 (276), same data.

NOTES: Solander's name Sciaena rubens does not appear to have been adopted by any later naturalist.

150.(**2**:63*b*) *Centracanthus cirrus* Rafinesque, 1810 Emmelichthyidae

DRAWING: finished water-colour, with ink detail of head with jaws extended; *r.* [ink] 'Sciaena angustata./Sydney Parkinson pinx[t]. 1768.'; *v.* [ink] 'Madeira'. 234 × 293.

MANUSCRIPT: Solander – (D. & W. 45) S.C. Pisces 2, f.27–29 as Sciaena angustata, 'Habitat in Oceano Atlantico prope insulam Maderam'; specimen 7 inches long. (D. & W. 42) C.S.D. f.245 (272), same data.

NOTES: the name *Sciaena angustata* was adopted by Valenciennes (1830) and published as *Smaris angustatus*, based on this drawing although Solander was credited as the author. This drawing therefore has some type standing.

Two separate specimens labelled *Smaris insidator* with no provenance were listed by Günther (1859) and might have been *Endeavour* specimens. Unfortunately they were both placed in the one bottle in the early twentieth century and that bottle was recently broken, so it is impossible to verify this conjecture.

151.(**2**:64*a*) *Lutjanus semicinctus* Quoy & Gaimard, 1824 Lutjanidae

DRAWING: unfinished water-colour; *r.* [pencil] 'Perca coregona mss/[ink] S. Parkinson'; *v.* [pencil] 'Sciena vittata/[ink] Princes Island'. 248 × 332.

MANUSCRIPT: Solander – not traced (see below and entry 152). Dryander –

Catalogue f. 155, under Sciaena, listed as Perca coregona Brouss. —— sketch with colours, Princes Island, S. Parkinson.

NOTES: there is no record of Sciaena vittata in the Solander manuscript Animalia Javanesis et Capensis (D. & W. 40d) in which it should have been included if Dryander's locality, Princes Island, is correct. Sciaena vittata is included in the manuscript describing the fishes and other animals from the Pacific Ocean (D. & W. 40c) – see number 152, this catalogue. Clearly the two drawings represent different taxa but the use of the same name is inexplicable.

Solander's name Sciena vittata seems never to have been employed by other naturalists although Lacepède (1802) published the same name apparently deriving it from Commerson's manuscript. The entry in Dryander's catalogue attributes the name Perca coregona to Broussonet but this also has not been taken up by later authors.

152.(**2**:64*b*) *Sargocentron diadema* (Lacepède, 1799) Holocentridae

DRAWING: unfinished water-colour; *r.* [pencil note by anal fin of fish] 'deep carmin/ [pencil] Ee-chi [ink] S. Parkinson'; *v.* [pencil] 'N°. 38 Sciena vittata/[ink] Otahite'. 256 × 329.

MANUSCRIPT: Solander – (D. & W. 40c) P.A.O.P. f. 24 (142), locality Tahiti, serial number A. 59 but not preserved (see f. 287). Dryander – Catalogue f. 155, possibly the drawing identified as Holocentrus macrophthalmus Brouss. —— Society Islands, S. Parkinson; otherwise any one of three *Sciaena* drawings from the same locality.

NOTES: Solander's name Sciena vittata seems not to have been used by later naturalists, but see discussion under number 151 in this catalogue.

153.(**2**:65) *Aplodactylus arctidens* Richardson, 1839 Aplodactylidae

DRAWING: unfinished water-colour; *r.* [pencil] 'Sciena Meandrites mss/Sciaena [ink] S. Parkinson'; *v.* [pencil] 'the whole back of this fish & the finns a Green gray Speckled wt black/The ground colour gradually turning pale towards the belly./ N°. 1 Meandrites/[ink] Mattaruwhow'. 298 × 470.

MANUSCRIPT: Solander – (D. & W. 40a) P.A. f. 2 (4) as Meandrites, Habitat prope Cape Kidnappers; f. 68 index entry only. Dryander – Catalogue f. 157 under pencil genus heading Meandrites sketch with colours, Sciaena Meandrites Brouss. —— New Zealand, S. Parkinson.

NOTES: Richardson (1842*b*) quoted Solander's manuscript account of Sciaena meandratus in the discussion of his description of the species *Aplodactylus arctidens*, opening his quotation with the comment that 'The first example of this genus was discovered by Solander on the coast of New Zealand, and named by him "Sciaena maeandratus." . . .'. Nowhere does he state or imply that this was a proposal of a new name. Despite this many authors have used the name *Sciaena maeandratus* as if it was proposed as a valid binomen, for example Whitehead (1968) who claims the Parkinson drawing was the basis for *Aplodactylus meandratus* Richardson, 1842.

However, Richardson (1843*a*) also quotes his earlier paper as if the Solander name was validly proposed; here too he refers to the drawing, as he does also in Richardson (1843*b*).

This drawing was reproduced by Whitehead (1968) as Plate 19.

154.(**2**:66) *Latridopsis ciliaris* (Bloch & Schneider, 1801) Latridae

DRAWING: pencil sketch; *r.* [ink] 'S. Parkinson'; *v.* [pencil] 'The upper side of the fish dark grey losing itself by degrees in a silvery colour the/finns & tail dark grey./ 19 Sciena salmonea/[ink] Totarranue'. 288 × 475.

MANUSCRIPT: Solander – not described but listed as Sciena salmonea at P.A. index (f.72) as two specimens (D. & W. 40a). Dryander – Catalogue f.155–7, presumably one of the seven drawings of Sciaena —— made by Parkinson at New Zealand.

NOTES: this species was described by Schneider from J. R. Forster's manuscript notes made on the *Resolution* voyage.

Solander's name Sciena salmonea does not appear to have been used by later naturalists.

This drawing was cited by Richardson (1842*b*) in his discussion of the previously undescribed genus *Latris*.

155.(**2**:67) *Arripis trutta* (Bloch & Schneider, 1801) Arripidae

DRAWING: unfinished water-colour; *r.* [pencil] 'Sciaena mulloides mss/[ink] S. Parkinson'; *v.* [pencil] '9. Mulloides sapidissimus/[ink] Opoorage'. 330 × 461.

MANUSCRIPT: Solander – (D. & W. 40a) P.A. f.17 (19), as Mulloides sapidissimus, habitat Tegadu, Tolaga, one specimen numbered B.26; also (f.68) index, where the serial number is not recorded. Dryander – Catalogue f.155 as Sciaena mulloides Brouss. —— sketch with colours, New Zealand, S. Parkinson.

NOTES: Solander's name Mulloides sapidissimus was published as a reference in synonymy by Richardson (1843*b*) who referred both to the manuscript and the Parkinson drawing under his name *Centropristes sapidissimus*. Richardson noted the differences in the localities cited by Solander and Parkinson. Richardson (1842*b*) also discussed the drawing and manuscript and quoted extensively from the latter.

This species was first named by Schneider from J. R. Forster's manuscript notes made on the *Resolution* voyage. Although Cuvier (1828) refers to G. A. Forster's drawing of *Sciaena trutta* in Banks's collection and had had the drawing copied he appeared to be unaware of the Parkinson drawing.

This drawing was reproduced by Whitehead (1963) at Plate 10.

156. (**2**:68) *Arripis trutta* (Bloch & Schneider, 1801) Arripidae

DRAWING: unfinished water-colour; *r.* [pencil] 'Sciaena mulloides./Hekāwai/ [ink] S. Parkinson'; *v.* [pencil] '9 Mulloides sapidissimus/[ink] Queen Charlotts Sound'. 298 × 279.

MANUSCRIPT: see number 155.

NOTES: Richardson (1842*b*) referred to this drawing within the discussion of *Centropristes salar* Richardson, 1839, but there is no evidence there that Richardson intended the name *Centropristes mulloides* to be read as a valid binominal (indeed the combination nowhere appears in that paper). Despite this some authors seem to have regarded it as the proposal of a new name including Richardson (1843*a*, *b*) himself. In the first of these two references Richardson cites this drawing by number.

157.(**2**:68–69) ? *Centropomus undecimalis* (Bloch, 1792) Centropomidae

DRAWING: unfinished water-colour; *r.* [pencil] 'Sciaena'; *v.* [pencil] 'N° 18. Gadoides/[ink] Brasil'. 268 × 374.

MANUSCRIPT: Solander – not traced. Dryander – not traced. See Notes below.

NOTES: the absence of a binominal name makes it impossible to trace this drawing in the manuscripts. In fact, Sciaena gadoides does occur in P.A. f.42 (44) (D. & W. 40a) but refers to a fish with Habitat off Motuaro, New Zealand. If the locality on the drawing is correctly attributed by Dryander this name and description cannot refer to this drawing.

The drawing is very imperfect; its style is uncertain and it is difficult to attribute it to an artist. If the locality, Brazil, is correct it is possibly by Buchan and in some respects it has similarities with his uncertain line and colouring. It may, however, be by Parkinson. It is noteworthy that Dryander did not certainly attribute it to any artist.

The identification of the drawing is uncertain. The Brazilian locality suggests *Centropomus*, but it is not an accurate representation of that species, and certain features suggest it might be a sciaenid.

158.(**2**:69) *Anisotremus surinamensis* (Bloch, 1791) Sparidae

DRAWING: finished water-colour; *r.* [pencil] 'Perca [written above as a substitute for Sciaena]/[ink] Sciaena labiata/S. Parkinson'; *v.* [ink] 'Brasil'. 295 × 462.

MANUSCRIPT: Solander – (D. & W. 45) S.C. Pisces 2, f.30–33, as Sciaena labiata, Habitat in Brasilia, length 20 inches; (D. & W. 42) C.S.D. 247 (274), same data. Dryander – Catalogue f.155 as Sciaena labiata mss, finished in colours, Brasil, S. Parkinson.

NOTES: Solander's name Sciaena labiata does not appear to have been employed by any later naturalist.

159.(**2**:70) *Sargocentron tiere* (Cuvier in Cuvier & Valenciennes, 1829) Holocentridae

DRAWING: unfinished water-colour; *r.* [pencil] 'holocentrus macrophthalmus/[ink] S. Parkinson'; *v.* [pencil] 'the whole fish of a bright carmine colour somewhat paler before as are the spines/the spots on the P.D. white the first circle of the eye purple 2nd red pupil black/N° 43 Sciaena rubra/[ink] Otahite'. 267 × 372.

MANUSCRIPT: Solander – (D. & W. 40c) P.A.O.P. f.24 (142) as Sciaena rubra, serial number A.58; (f.287) index and list of specimens collected, suggests that no specimen was preserved. Dryander – Catalogue f.155, originally as Sciaena but with Holocentrus macrophthalmus Brouss. pencilled in, sketch with colours, Society Islands, S. Parkinson.

NOTES: the name Sciaena macropthalma occurs in Solander's manuscript P.A.O.P. at f.9 (123), and the entry is cross-referred to that for Sciaena rubra. It seems from the evidence of Dryander's Catalogue that Broussonet considered them identical and wrote the name S. macropthalma on the recto of the drawing. Neither name appears to have been used from Solander's manuscript by later naturalists, although when *Sciaena rubra* Bloch & Schneider, 1801, was used it was an independent proposal.

160.(**2**:71) *Pterocaesio tile* (Cuvier in Cuvier & Valenciennes, 1830) Lutjanidae

DRAWING: unfinished water-colour; *r.* [pencil] 'Sciaena melanura mscr./Aaur-èoorè/[ink] S. Parkinson'; *v.* [pencil] 'N°. 40. Laveratoides amaenus/[ink] Otahite'. 271. × 373.

MANUSCRIPT: Solander – (D. & W. 40c) P.A.O.P. f.48 (166), as Lavaretoides amaenus, vernacular name, one specimen 'unmark'd' in Cagg 6. Dryander – Catalogue f.155–6 not identified by name amongst the *Sciaena* species; Lavaretoides is not entered in the catalogue.

NOTES: Solander's name Lavaretoides amaenus does not appear to have been employed by later naturalists, nor does Sciaena melanura which is believed to be a Broussonet name. Cuvier (1830) in describing the species *Caesio tricolor* refers to a Parkinson drawing in the Banks collection captioned *Sciaena*. This is the only *Caesio* (s.l.) drawing in the collection and may be the one referred to, although the description of the coloration does not entirely agree (however, it has to be stressed that Cuvier was working from a copy of the drawing, not the original).

161.(**2**:72) *Pagrosomus auratus* (Houttuyn, 1782) Sparidae

DRAWING: unfinished water-colour; *r.* [pencil] 'Sparus erythrinus Linn ? [ink] S. Parkinson'; *v.* [pencil] 'on the back ar a number of silvery spots with a blue cast./6 Sciena lata/[ink] Oahoorage'. 290 × 475.

MANUSCRIPT: Solander – (D. & W. 40a) P.A. f.22 (24), as Sciena lata, habitat coast near Opuragi and Oouhuragi (Mercury Bay and Hauraki Gulf). Dryander – Catalogue, not identifiable amongst the two *Sparus* and four *Sciaena* drawings at ff.149 and 155.

NOTES: Solander's description was quoted extensively by Richardson (1842*a*) in establishing his species *Pagrus latus*; the Parkinson drawing was also discussed and has standing as type material. Richardson also referred to G. Forster's drawing labelled *Sciaena aurata*, made on the *Resolution* voyage, the description of which formed the basis of *Labrus auratus* Bloch & Schneider, 1801 (Wheeler, 1981). The identification written on the recto of the drawing *Sparus erythrinus* Linn ? was probably written by Broussonet who associated this drawing with this European seabream.

This drawing is reproduced by Whitehead (1968) as Plate 13.

162.(**2**:73) *Plectorhynchus picus* (Cuvier in Cuvier & Valenciennes, 1830) Pomadasyidae

DRAWING: unfinished water-colour; *r.* [pencil] 'Perca scelerata ms/Abootoo/ [ink] S. Parkinson'; *v.* [pencil] 'N°. 15 Labrus punctatus/ [ink] Otahite'. 268 × 373.

MANUSCRIPT: Solander – (D. & W. 40c) P.A.O.P. f.44 (162) as Labrus punctatus; (f.289) no specimen preserved. Dryander – Catalogue f.151 as *Labrus*, pencil addition Perca scelerata Brouss. —— sketch with colours, Society Islands, S. Parkinson.

NOTES: the Solander name Labrus punctatus seems not to have been used by later naturalists, probably because it was preoccupied by *Labrus punctatus* Linnaeus, 1758. Perca scelerata, written on the recto of the drawing, is one of Broussonet's numerous annotations.

The drawing which Cuvier referred to in his description of *Diagramma pica* is listed at number 167 in this catalogue.

163.(**2**:74) *Polyprion oxygenios* (Bloch & Schneider, 1801) Percichthyidae

DRAWING: unfinished water-colour; *r.* [pencil] 'Perca Gadoides mss./ [ink] S. Parkinson'; *v.* [pencil] '13. Sciaena gadoides/ [ink] Motuaro'. 298 × 489.

MANUSCRIPT: Solander – (D. & W. 40a) P.A. f.42 (44) as Sciaena gadoides, habitat off Motuaro, no serial number, not preserved (f.72), total length 27 inches. Dryander – Catalogue f.157 as Sciaena, pencil addition Perca gadoides Brouss. —— sketch with colours, New Zealand, S. Parkinson.

NOTES: *Epinephelus oxygenios* was proposed by Schneider (1801) from the description made by J. R. Forster on the *Resolution* voyage of a specimen from Queen Charlotte's Sound, New Zealand. However, Schneider did not see the drawing which was in Banks's collection in London (Wheeler, 1981). This drawing was copied by Mrs S. Bowdich and Cuvier (1829) referred to it, but made no reference to the Parkinson drawing of the same taxon. Neither Solander's name Sciaena gadoides nor Broussonet's amendation to Perca gadoides seem to have been used by later naturalists.

164.(**2**:75*a*) *Lutjanus kasmira* (Forsskål, 1775) Lutjanidae

DRAWING: unfinished water-colour; *r.* [pencil] 'Eta äpà/ [ink] S. Parkinson'; *v.* [pencil] 'There is/some of this fish much larger/N°. 39. Perca vittata/ [ink] Otahite'. 249 × 344.

MANUSCRIPT: Solander – (D. & W. 40c) P.A.O.P. f.16 (134) as Perca vittata, several specimens, perhaps 5 in Cagg 4 numbered A.38, and two 'unmark'd' in Cagg 6. Dryander – Catalogue not identified by species, one of several *Perca* from the Society Islands drawn by Parkinson.

NOTES: Solander's name Perca vittata seems not to have been adopted by later naturalists. There are two specimens listed by Günther (1859) under *Genyroge bengalensis* specimens d, and e, 'Adult India Old Collection'; they might be from the *Endeavour* voyage, but are as likely to be from the *Resolution* voyage (see Wheeler, 1981).

165.(**2**:75*b*) *Epinephelus merra* Bloch, 1793 Serranidae

DRAWING: finished water-colour; *r*. [ink] 'Serranus hexagonatus/ [pencil] Etaràaò op appah/ [ink] S. Parkinson'; *v*. [pencil] 'Tarão The ground colour & the spots are darker & soften'd in to one another round about each spot/are small dots of white or straw colour – the same across the finns. there is of this fish as large again/N°. 36. Perca maculata/ [ink] Otahite'. 244 × 356.

MANUSCRIPT: Solander – (D. & W. 40c) P.A.O.P. f.30 (148) as Perca maculata, three specimens two numbered A.65, and one A.127; f.288 confirms that three specimens were preserved. Dryander – Catalogue ff.157–8 presumed to be one of several entries under *Perca* —— of drawings made at Tahiti by Sydney Parkinson.

NOTES: this drawing was the basis of Valenciennes's (1829) *Serranus Parkinsonii* the name and description being derived from the drawing alone. The name *Perca maculata* was used by J. R. Forster in his manuscript account of the *Resolution* animals (Forster, 1844) but applies to the species *Epinephelus fasciatus* (Forsskål, 1775), see Wheeler (1981) for discussion. There are two specimens in the British Museum (Natural History) which may be *Resolution* specimens, and a third in the Muséum National d'Histoire Naturelle, Paris, received by Cuvier from Broussonet and originally coming from Banks (Wheeler, 1981; Bauchot, 1969). The Paris specimen and the London specimens might have come from either the *Endeavour* collection or the *Resolution*, or both.

This drawing was reproduced by Whitehead (1968) as Plate 12.

166.(**2**:76) *Paracirrhites arcatus* (Cuvier in Cuvier & Valenciennes, 1829) Cirrhitidae

DRAWING: unfinished water-colour; *r*. [ink] 'S. Parkinson'; *v*. [pencil] 'N°. 26. Perca areata/ [ink] Otahite'. 268 × 370.

MANUSCRIPT: Solander – (D. & W. 40c) f.27 (145), as Labrus areatas, one specimen serial number A.64; f.107 (227) as Perca areata, two specimens with serial number A.167; also indexed f.288. Dryander – Catalogue ff.158–9 not identified by name amongst the *Perca* —— entries for drawings from the Society Islands by Parkinson.

NOTES: Cuvier (1829) based his name *Cirrhites arcatus* partly on this Parkinson drawing, but had a specimen and another drawing as well. He derived the species name from the copy of the drawing provided by Mrs S. Bowdich, who had erroneously copied the Solander name 'Perca areata' as *Perca arcata*. This discrepancy was noted by Richardson (1848) who quoted extensively from the Solander manuscript citing both accounts and differentiating by their serial numbers (A.64 and A.167).

Fig. 11 *Paracirrhites arcatus* (Cuvier *in* Cuvier & Valenciennes, 1829). Parkinson's drawing from a Tahitian fish was used as part basis by Cuvier for the description of this species, although because of a copyist's error the trivial name was changed from Solander's areata to *arcata*. (Catalogue number 166.)

167.(**2**:77) *Plectorhynchus picus* (Cuvier in Cuvier & Valenciennes, 1830)
Pomadasyidae

DRAWING: unfinished water-colour; *r.* [pencil] 'Tairhepha/ [ink] S. Parkinson'; *v.* [pencil] 'The parts mark'd thus x are white inclining to gray especially on the finns/ & on the face reddish. those mark'd w^{t} 2 are black the scales edged w^{t} dirty white./ The iris gold colour pupil black./N°. 45 Percoides pica/ [ink] Otahite'. 268 × 372.

MANUSCRIPT: Solander – (D. & W. 40c) P.A.O.P. f.19 (137) as Percoides Pica, one specimen serial number A.39; (f.287) index. Dryander – Catalogue – not traced, the name Percoides is not entered, and there is no entry Perca pica.

NOTES: Solander's name pica was employed by Cuvier (1830) in the form *Diagramma pica* and he referred to this drawing in his discussion of the species. He also had a specimen in his collection.

Another drawing of this species is listed at number 162 in this catalogue.

168.(**2**:78) *Cynoscion* sp. Sciaenidae

DRAWING: finished water-colour; *r.* [ink] 'S. Parkinson'; *v.* [pencil] 'N°. 10. Perca/ [ink] Brasil'. 297 × 478.

MANUSCRIPT: Solander – not traced. Dryander – Catalogue f.157 probably as *Perca* —— Brasil, S. Parkinson.

NOTES: as there is no trivial name on this drawing it is impossible to be certain to which entry in Dryander's Catalogue this drawing relates. The only Parkinson drawing from Brazil is said to be a 'sketch with colours' but this appears to be a finished drawing. On the other hand the detail of this drawing is dissimilar to Parkinson's usual style and it is possible that it is a Buchan drawing misattributed by Dryander in his annotation to the drawing. There are three Buchan drawings listed as *Perca* —— from Brazil in Dryander's Catalogue.

Like several other Brazilian fishes there is no entry for this in Solander's manuscript.

169.(**2**:79*a*) *Anthias anthias* (Linnaeus, 1758) Serranidae

DRAWING: finished water-colour; *r.* [ink] 'Perca Imperator./Sydney Parkinson. pinx[t] 1768./S. Parkinson'; *v.* [ink] 'Madeira'. 263 × 329.

MANUSCRIPT: Solander – (D. & W. 45) S.C. Pisces 2 f.67–70*v*, as Perca Imperator, habitat in Oceano Atlantico Maderae, total length 7½ inches. (D. & W. 42) C.S.D. f.257 (276), same data.

NOTES: Solander's name Perca Imperator (derived from his note of the Madeiran vernacular 'Emperador') seems not to have been employed by later naturalists.

There is a specimen in the collection of the British Museum (Natural History) from Madeira, in the Old Collection as *Perca imperator* (see Günther, 1859) which is undoubtedly the *Endeavour* specimen.

170.(**2**:79*b*) *Anthias anthias* (Linnaeus, 1758) Serranidae

DRAWING: unfinished water-colour, pencil details of head and pelvic fin; *r.* [ink] 'S. Parkinson Perca imperator T: 17 Madeira'; *v.* [pencil] 'Mem. that the P.A. is somewhat brownish & the Pinna V. is upon the orange lay the spots upon the head/ of a very delicate scarlet N.B. M[r] B. thinks it too pale'. 213 × 292.

MANUSCRIPT: see above number 169.

NOTES: this must have been a preliminary sketch of this fish which was redrawn to take account of Banks's comments recorded in the annotation. See also number 169.

171.(**2**:80) *Mycteroperca rubra* (Bloch, 1793) Serranidae

DRAWING: finished water-colour by A. Buchan; *r.* [ink] 'Buchan'; *v.* [pencil] 'N°. 11. Perca/[ink] Brasil'. 269 × 368.

MANUSCRIPT: Solander – not traced; there are no entries under the genus *Perca* in the Slip Catalogue (D. & W. 42). Dryander – Catalogue f.157 presumed to be one of the three Brasilian drawings of *Perca* —— by A. Buchan.

NOTES: there are very few entries for Brazilian fishes in Solander's manuscripts; this is another example where no description was made.

Fig. 12 *Mycteroperca rubra* (Bloch, 1793). Drawing by Buchan of a Brazilian fish. (Catalogue number 171.)

172.(**2**:81) *Cephalopholis urodelus* (Bloch & Schneider, 1801) Serranidae

DRAWING: unfinished water-colour; *r.* [ink] 'S. Parkinson [pencil] Matapoo'; *v.* [pencil] 'The whole body is dark red especially toward the tail & back./spotted w[t] bright scarlet./N°. 37. Perca escarlatina/[ink] Otahite'. 269 × 373.

MANUSCRIPT: Solander – (D. & W. 40c) P.A.O.P. f.23 (141) as Perca escarlatina, one specimen serial number A.56; f.34 (152) one specimen, number A.68; f.35 (153) two specimens A.72, and one A.190; f.(287) lists four of these in the index and list of specimens. Dryander – Catalogue f.157–8, not identified amongst the seven *Perca* —— drawings by Parkinson from the Society Islands.

NOTES: the name Perca escarlatina appears not to have been adopted by any later naturalist.

173.(**2**:82) *Mycteroperca* sp. Serranidae

DRAWING: finished water-colour; *r.* [ink] 'Perca. asellina./Sydney Parkinson pinx[t] 1769'; *v.* [pencil] '16 Perca/Gadoides/[pencil – four lines of unreadable notes]/ [ink] Brasil'. 294 × 478.

MANUSCRIPT: Solander – (D. & W. 45) S.C. Pisces 2, f.57–58v, as Perca asellina, Habitat at Rio Janeiro; (D. & W. 41) f.251 (268), same data. Dryander – Catalogue f. 157 as Perca asellina mss, finished in colours —— Brasil, S. Parkinson.

NOTES: Solander's name Perca asellina seems not to have been used by later workers.

The drawing appears to be of a large fish, but it is not possible to confirm this as for once no length was given for the specimen in Solander's description. The inferred large size of the specimen has led to an apparently inaccurate drawing which is difficult to identify.

174.(**2**:83) *Serranus atricauda* Günther, 1874 Serranidae

DRAWING: pencil & wash; *r.* [ink] 'Perca – decorata/Sydney Parkinson pinx[t] 1768'; *v.* [ink] 'Madeira'. 270 × 373.

MANUSCRIPT: Solander – (D. & W. 45) S.C. Pisces 2, f.63–64v, as Perca decorata, 'Habitat in Oceano Atlantico prope Maderam', length of the specimen 10 inches; (D. & W. 41) f.255 (272), same data. Dryander – Catalogue f. 157 as Perca decorata, finished without colour, —— Madeira, S. Parkinson.

NOTES: the *Endeavour* specimen is still preserved in the British Museum (Natural History). It was listed by Günther (1859) as *Serranus cabrilla*, specimen i. 'Adult: not good state. Madeira. Old Collection as *Perca decorata*'. Boulenger (1895) listed it with several other specimens which Günther had originally called *S. cabrilla* as one of the 'Types' of *Serranus atricauda*, but it is doubtful whether it should be regarded as a type because Günther (1874) had merely described the species from a specimen from Morocco and commented that it was 'identical with others in the British Museum from the Azores, Madeira and the Canary Islands'. The *Endeavour* fish was one of the Madeiran specimens.

Other than being listed in synonymy by Günther (1874) the Solander name Perca decorata has not been employed by later naturalists.

175.(**2**:84) *Epinephelus itajara* (Lichtenstein, 1822) Serranidae

DRAWING: finished water-colour by A. Buchan; *r.* [ink] 'Buchan/[pencil] Light red' [beside mouth]/[numerals beside fins indicating numbers of rays]; *v.* [pencil] 'N°. 9. Perca nebulosa/[ink] Brasil'. 268 × 370.

MANUSCRIPT: Solander – (D. & W. 45) S.C. Pisces 2, f.59–61v as Perca nebulosa, habitat in Oceano prope Fluvium Januarii Brasiliae [Rio de Janeiro]; (D. & W. 42) C.S.D. f.253 (270), same data. Dryander – Catalogue f. 157, not listed by name but presumed to be one of three *Perca* —— drawings by Buchan from Brazil.

NOTES: the Solander name Perca nebulosa seems never to have been used by later naturalists.

176.(**2**:85) *Micropogon ? undulatus* (Linnaeus, 1766) Sciaenidae

DRAWING: finished water-colour by A. Buchan; *r.* [pencil] 'Perca undulata L. ?/

[ink] A. Buchan'; *v.* [pencil] 'N°. 12. Perca/[ink] Brasil'. 268 × 369.

MANUSCRIPT: Solander – not traced. Dryander – Catalogue f.157 presumed to be one of three *Perca* —— drawings by Buchan from Brazil.

NOTES: as with several other fishes caught and drawn at Brazil there seems to be no description by Solander of this specimen which served as a basis for Buchan's drawing.

177.(**2**:86) *Naucrates ductor* (Linnaeus, 1758) Carangidae

DRAWING: finished water-colour; *r.* [ink] 'Gasterosteus – Ductor/Sydney Parkinson pinx[t] 1768'; *v.* [none]. 224 × 292.

MANUSCRIPT: Solander – (D. & W. 45) S.C. Pisces 2, f.88 and (D. & W. 42) C.S.D. f.88, both as *Gasterosteus Ductor* with reference to Linnaeus and meristic data but no habitat given or description. Dryander – Catalogue f.161 as *G. Ductor* L., finished with colour, Ocean, S. Parkinson.

NOTES: the pilot fish was well known to sailors and was described by Linnaeus from the accounts and specimens of other travellers. Presumably Solander did not consider it necessary to describe this species in detail.

178.(**2**:87) *Acanthocybium solandri* (Cuvier in Cuvier & Valenciennes, 1832) Scombridae

DRAWING: finished water-colour; *r.* [ink] 'Scomber lanceolatus/Sydney Parkinson pinx[t] 1769/[pencil] Tatea/[ink] Mem. one Pinulae spuriae is wanting above & one below'; *v.* [pencil] 'off Thrum Cap. Island'. 297 × 462.

MANUSCRIPT: Solander – (D. & W. 45) S.C. Pisces 2, f.107–110v as Scomber lanceolatus, 'habitat in Oceano australi seu Mari Pacifico, prope Insulas ("Thrum Cap"), Apr. 4, 1769 . . .', length 4 feet; (D. & W. 42) C.S.D. f.267 (284), same data. Dryander – Catalogue f.163 as finished in colour, Scomber lanceolatus Mss —— Ocean, S. Parkinson.

NOTES: Solander's manuscript was the sole source of Cuvier's information for his description and name *Cybium Solandri*. He made no reference to this drawing which is undoubtedly of the specimen Solander described. No locality was given by Cuvier (1832), and Günther (1860) likewise questioned the geographical origin of the specimen. This uncertainty has continued till the present day because Collette & Nauen (1983) comment that the locality was unknown. Thrum Cap Island lies immediately south of the island of Hau in the Tuamotu archipelago, and this is the type locality for this species.

179.(**2**:88) *Selar crumenophthalmus* (Bloch, 1793) Carangidae

DRAWING: unfinished water-colour; *r.* [ink] 'S. Parkinson/[ink] Etoore/Owhey'; *v.* [pencil] 'N°. 41. Scomber albula/[ink] Otahite'. 265 × 372.

MANUSCRIPT: Solander – (D. & W. 40c) P.A.O.P. f.10 (124) as Scomber albula, a total of four specimens preserved, one numbered A.26 'magna', three of medium size numbered A.73, from George Land (= Tahiti). Dryander – Catalogue f.161–2, not entered as Scomber albula, presumably one of the three *Scomber* —— drawings from the Society Islands, by S. Parkinson,

NOTES: Solander's name Scomber albula seems never to have been employed by later naturalists.

180.(**2**:89) *Caranx lutescens* (Richardson, 1843) Carangidae

DRAWING: unfinished water-colour; *r.* [ink] 'S. Parkinson'; *v.* [pencil] 'the belly an opaline colour – /12. Scomber micans/ [ink] Opoorage'. 292 × 473.

MANUSCRIPT: Solander – (D. & W. 40a) P.A. f.33 (35) as Scomber micans, habitat off Motuaro. Dryander – Catalogue f.163, one of two drawings of *Scomber* —— sketch with colours, New Zealand, S. Parkinson.

NOTES: Solander's name Scomber micans was published by Richardson (1843*c*) and his description was quoted extensively in a discussion of carangid fishes known from New Zealand. Parkinson's drawing was also referred to. Both the description and figure were referred to elsewhere by Richardson (1843*b*) in the synonymy of *Caranx georgianus*, as they were by Richardson (1848) under the same name but with discussion.

Scomber lutescens was based by Richardson (1843) on Solander's manuscript description of a specimen caught in New Zealand waters on 30 March 1770 in Queen Charlotte Sound, but not illustrated (see D. & W. 40a, P.A. f.51 (53)).

181.(**2**:90) *Pomatomus saltatrix* (Linnaeus, 1766) Pomatomidae

DRAWING: finished water-colour; *r.* [ink] 'Scomber – salmoneus. /Sydney Parkinson – pinx^t 1769./ [pencil] gasterosteus saltatrix Linn'; *v.* [pencil – two lines of indecipherable notes/and name] / [ink] 'Brasil'. 298 × 465.

MANUSCRIPT: Solander – (D. & W. 45) S.C. Pisces 2, f.125–125v as Scomber saltatrix [specific epithet substituted for salmoneus], habitat in Oceano Brasiliensis; (D. & W. 42) C.S.D. f.277 (145), same data. Dryander – Catalogue f.161 as Scomber salmoneus mss, finished in colour —— Brasil, S. Parkinson.

NOTES: *Pomatomus saltatrix* was described by Linnaeus in 1766 (as *Gasterosteus*) on the basis of Catesby's earlier description and a specimen in the Garden collection (Wheeler, 1985). Possibly Solander failed to recognize the Linnaean species because of its obvious affinity to the genus *Scomber*, not *Gasterosteus*, and therefore proposed the specific epithet salmoneus. However, he later recognized it as identical with Linnaeus's species and amended his manuscript, and the name on the drawing was altered also.

182.(**2**:91) *Rexea solandri* (Cuvier in Cuvier & Valenciennes, 1832) Gempylidae

DRAWING: unfinished water-colour; *r.* [ink] 'S. Parkinson/ [in pencil on mount] Gempylus solandri *Cuv.* & *Val.*'; *v.* [pencil] '15 Scomber macrophthalmus/ [ink] Aehie no Mauwe'. 296 × 469.

MANUSCRIPT: Solander – (D. & W. 40a) P.A. f.44 (46) as Scomber macrophthalmus, habitat Oceano australium, 9 December 1769; f.72 listed in index. Dryander – Catalogue f.163, presumed to be one of the two *Scomber* —— sketches with colours, drawn in New Zealand by S. Parkinson.

NOTES: Solander's name Scomber macrophthalmus was quoted by Cuvier (1832) when he described *Gempylus solandri*. This species was based solely on Solander's manuscript account and Cuvier made no reference to the Parkinson drawing. Possibly it had not been copied for him. He also claimed that the fish came from 'la mer de la Nouvelle-Holland', presumably confused by the title of Solander's manuscript 'Pisces Australiae'.

On 9 December 1769 the *Endeavour* had just left the Bay of Islands en route for the South Island of New Zealand.

Reproduced by Whitehead (1968) as his Plate 21 and alleged to be the basis for the species *Gempylus solandri* although as noted above Cuvier made no reference to the drawing.

183.(**2**:92) *Gempylus serpens* Cuvier, 1829 Gempylidae

DRAWING: finished water-colour; *r.* [ink] 'Scomber-serpens/Sydney Parkinson pinx^t^ 1768'; *v.* [ink] 'Septr. 23. 1768/of Canary Islands'. 299 × 469.

MANUSCRIPT: Solander – (D. & W. 45) S.C. Pisces 2, f.111–114*v* as Scomber serpens, 'habitat in Oceano Atlantico prope Insulas canariensis (Septr. 22. 1768 captus.)', length 37 inches; (D. & W. 42) C.S.D. f.269 (137), same data. Dryander – Catalogue f.161 as Scomber serpens Mss —— Ocean, S. Parkinson.

NOTES: Cuvier (1832) referred to Solander's description of Scomber serpens and quoted from it in part, but he also had a dry specimen from the Caribbean ('Antilles') to refer to. The first record and figure of this rare oceanic fish was made by Hans Sloane on his voyage to Jamaica (Sloane, 1707) and this was also referred to by Cuvier.

184.(**2**:93) *Caranx melampygus* Cuvier in Cuvier & Valenciennes, 1833 Carangidae

DRAWING: unfinished water-colour; *r.* [pencil] 'Eürùā/Eppouea/[ink] S. Parkinson'; *v.* [pencil] 'all spotted but the belly, fins & tail a dirty grey, about the bottom of the tail a list of dark blue/N°. 3. Scomber stellaris/[ink] Otahite'. 295 × 467.

MANUSCRIPT: Solander – (D. & W. 40c) P.A.O.P. f.104 (224), possibly as Scomber stellatus, length 18 inches (but see number 186 this catalogue). Dryander – Catalogue ff.161–163, presumed to be one of the four *Scomber* —— drawings made by Parkinson in the Society Islands but not identified to species in the catalogue of drawings.

NOTES: Solander's name, Scomber stellaris (as given on the drawing) or S. stellatus (as in his manuscript) does not seem to have been adopted by later workers. Cuvier (1833) in proposing the name *Caranx melampygus* made no reference to either Parkinson's drawing or Solander's description.

185.(**2**:94) *Carangoides crysos* (Mitchill, 1815) Carangidae

DRAWING: finished water-colour; *r.* [ink] 'Scomber falcatus./Sydney Parkinson pinx[t] 1768'; *v.* [ink] 'Nov[r]. 8th. 1768/Coast of Brasil'. 299 × 460.

MANUSCRIPT: Solander – (D. & W. 45) S.C. Pisces 2, f.115–118v as Scomber falcatus, 'habitat in Oceano Brasiliam alluente', length 16 inches; (D. & W. 42) C.S.D. f.271 (139), same data. Dryander – Catalogue f.161, as Scomber falcatus mss —— Brasil, S. Parkinson.

NOTES: Solander's name Scomber falcatus has not been adopted by later workers.

186.(**2**:95) *Caranx melampygus* Cuvier in Cuvier & Valenciennes, 1833 Carangidae

DRAWING: unfinished water-colour; *r.* [ink] 'S. Parkinson [pencil] Owrooå'; *v.* 'Mem. the back & part of the sides are spotted w[t] blue, the blue should be Ultramarine/the belly opaline./N°. 64 Scomber stellatus/[ink] Otahite'. 267 × 370.

MANUSCRIPT: Solander – (D. & W. 40c) P.A.O.P. f.104 (224) as Scomber stellatus, see number 184 in this catalogue.

NOTES: see above number 184. It is not possible to know whether the two drawings labelled Scomber stellaris and S. stellatus, were both intended to refer to the single description of S. stellatus, or whether S. stellaris was omitted from the manuscript in error.

187.(**2**:96) *Katsuwonus pelamis* (Linnaeus, 1758) Scombridae

DRAWING: unfinished, pencil and ink; *r.* [ink] 'S. Parkinson'; *v.* [pencil] 'N°. 1 Scomber Pelamys'. 291 × 462.

MANUSCRIPT: Solander – (D. & W. 45) S.C. Pisces 2, ff.95–100, as Scomber Pelamis; (D. & W. 42) C.S.D. f.263 (280), same data. Dryander – Catalogue f.161 as sketch without colours, Scomber pelamis L., Ocean, S. Parkinson.

NOTES: this well-known tuna was described in great detail by Solander who recognized it as the Linnaean species. The specimen on dissection contained two internal parasites which Solander described in manuscript as Fasciolis Pelamini and Sipunculus Piscium.

188.(**2**:97) *Oligoplites saurus* (Bloch & Schneider, 1801) Carangidae

DRAWING: unfinished water-colour; *r.* [ink] 'S. Parkinson/[pencil] Scomber saurus mss.'; *v.* [pencil] 'Mem. the Belly is like Silver the rest of the fins are grey./ N°. 8 Scomber/[ink] Brasil'. 269 × 373.

MANUSCRIPT: Solander – (D. & W. 45) S.C. Pisces 2, f.120 as Scomber saurus. Dryander – Catalogue f.161 as sketch with colours, Scomber saurus Brouss. —— Brasil, S. Parkinson.

NOTES: the Solander manuscript entry in the Slip Catalogue (see above) was written after the *Endeavour* voyage and evidently described a specimen sent to Banks by Roger Shakespear around 1779 (Dawson, 1958), but the specific epithet derives from Patrick Browne's (1756) *The civil and natural history of Jamaica* in which he refers to this species as Saurus number 1. This explains the coincidence of Bloch (in Bloch & Schnieider, 1801) adopting the same specific epithet. To the Solander slip a pencil addition has been made, probably by Broussonet, referring to the Parkinson drawing as 'Scomber N°. 8 Brasil'. This presumably was the occasion when the drawing was annotated as noted above, and this was the reason Dryander attributed the name to Broussonet.

189.(**2**:98) *Scomberoides lysan* (Forsskål, 1775) Carangidae

DRAWING: unfinished water-colour; *r.* [pencil] 'Scomb. Glaucus L/Eràì/[ink] S. Parkinson'; *v.* [pencil] 'the belly silvery/N°. 14. Scomber laevis/[ink] Otahite'. 265 × 374.

MANUSCRIPT: Solander – (D. & W. 40c) P.A.O.P. f.16 (134) as Scomber laevis, length 16½ inches. Dryander – Catalogue f.161–3, not identified by name, presumed to be one of three Parkinson drawings from the Society Islands listed as *Scomber*.

NOTES: Solander's name Scomber laevis does not appear to have been used by later naturalists. The annotation Scomb. Glaucus L. on the recto of the drawing was probably written by Broussonet under the misapprehension that this fish was identical with Linnaeus's species of that name now regarded as a junior synonym of *Trachinotus ovatus* (Linnaeus, 1758), see Wheeler (1963) for discussion.

190.(**2**:99) *Seriola zonata* (Mitchill, 1815) Carangidae

DRAWING: finished water-colour; *r.* [ink] 'Scomber – amia./Sydney Parkinson pinx[t] 1768'; *v.* [ink] 'Nov[r]. 8. 1768/Coast of Brasil'. 290 × 462.

MANUSCRIPT: Solander – (D. & W. 45) S.C. Pisces 2, f.121–124*v* as Scomber amia L., Habitat in Oceano Brasiliano, total length 39 inches; (D. & W. 42) C.S.D. f.275 (143), same data. Dryander – Catalogue f.161 as Scomber amia L. —— Brasil, S. Parkinson, finished in colours.

NOTES: Solander identified this fish with Linnaeus's *Scomber amia*, now known as *Lichia amia* (Linnaeus, 1758).

191.(**2**:100) *Thunnus albacares* (Bonnaterre, 1788) Scombridae

DRAWING: unfinished water-colour; *r.* [pencil] 'Eahè/aahei/[ink] S. Parkinson'; *v.* [pencil] 'The belly lead colour with an opal cast streakt & spotted w[t] silver the under part of the head silvery the P.P. lead colour/the P.D. & P.A. bright yellow the Iris silver the pupil black/76. Scomber Thynnus/[ink] August. 14 1769/off the Island of Oheteroa'. 292 × 471.

MANUSCRIPT: Solander – (D. & W. 40c) P.A.O.P. f.135 (255) as Scomber Thynnus [deleted by red and black vertical lines], habitat in oceano non procus ab insula Ohitirhoa (August 13, 1769). Dryander – Catalogue f. 161, as sketch with colours, *Scomber Thynnus* L. —— Ocean, S. Parkinson.

NOTES: Solander identified this tuna with the Linnaean species *Scomber thynnus*, which was the only large species known at the time (*Katsuwonus pelamis* (Linnaeus, 1758) being the comparatively small skipjack tuna). This specimen was captured soon after leaving the Society Islands on 9 August 1769.

192.(**2**:101) *Sarda sarda* (Bloch, 1793) Scombridae

DRAWING: finished water-colour by A. Buchan; *r.* [ink] 'Buchan [pencil] 22.16'; *v.* [ink] 'Brasil'. 265 × 370.

MANUSCRIPT: Solander – (D. & W. 45) S.C. Pisces 2, f. 100 as Scomber Pelamis 'varietas capta in Ostrio Fluvii Januarii (Rio de Janeiro) . . .'; (D. & W. 42) C.S.D. f. 263 (280), same data (see notes). Dryander – Catalogue f. 161 as *Scomber*, finished in colour —— Brasil, Buchan.

NOTES: the identification of this specimen in the Dryander Catalogue is certain as there is only one Buchan drawing of *Scomber*. Solander's notes imply that he examined two specimens which he identified as *Scomber pelamis* Linnaeus, 1758. In the *Slip Catalogue* there is a long and detailed description of a tunny occupying ff. 95–99*v* with, on f. 100, a note (as quoted above) referring to a variety of the species. This appears to have been written on a separate occasion to the main entry and probably refers to the fish Buchan drew, Solander apparently having considered it to be a variety of *Scomber pelamis* – it was not recognized as distinct and formally named for several years after the *Endeavour* voyage.

193.(**2**:102*a*) *Upeneus vittatus* (Forsskål, 1775) Mullidae

DRAWING: unfinished water-colour; *r.* [pencil] 'Eraòu ă/[ink] S. Parkinson'; *v.* [pencil] 'there is some of this 3 times as large/Èhuwīa/N° 19 Mugil [changed to] Mullus vitatus/[ink] Otahite'. 239 × 296.

MANUSCRIPT: Solander – (D. & W. 40c) P.A.O.P. f. 3 (115) as Mullus vittatus Fig. Pict. specimen numbered A9 from George Land, and f. 38 (156) two specimens numbered A75 and A159 in Caggs number 3 and 6 respectively. Dryander – Catalogue f. 173, as sketch with colours, *Mugil* —— Society Island, S. Parkinson.

NOTES: Solander's index to P.A.O.P. (f. 285) shows that he had preserved a total of six specimens of his Mullus vittatus, but the main text shows that the first described (at f. 3 (115)) was the one drawn. There are no specimens listed by Günther (1859) which could be *Endeavour* fishes, but there is a specimen in the Muséum National d'Histoire Naturelle, Paris, from Broussonet's collection (A 3461; S.L. 105, T.L. 137 mm) which is probably an *Endeavour* specimen (Bauchot, 1969). George Land was the name briefly in use on the expedition for Tahiti, although later usage was Otaheite.

194.(**2**:102*b*) *Prionotus* sp. Triglidae

DRAWING: pencil sketch; *r.* [ink] 'S. Parkinson'; *v.* [pencil] 'the body of a Greenish fusca spotted w^t^ redish brown the head has more of the yellow or orange init the Belly/a shell colour the P.P. nutmeg colour clouded near the upper side w^t^ black the under side edg'd w^t^ blue the P.V./has a tinge of red the P.D. grey spotted with reddish brown – the tail at the base the same colour as the belly/but the most part red spotted w^t^ dark brown – the Iris of the eye yellow pupil black. – N° 17. Trigla/[ink] Brasil'. 235 × 294.

MANUSCRIPT: Solander – not traced. Dryander – Catalogue f. 165 as *Trigla* ——, sketch without colours, Brasil, S. Parkinson.

NOTES: this appears to be another of the Brazilian fishes which were not described although they were drawn.

195.(**2**:103) *Dactylopterus volitans* (Linnaeus, 1758) Dactylopteridae

DRAWING: unfinished water-colour; *r.* [ink] 'S. Parkinson'; *v.* [pencil] 'The blue on the P.P. should be ultramarine/N° 6. Trigla volitans/[ink] Brasil'. 295 × 480.

MANUSCRIPT: Solander – not traced. Dryander – Catalogue f. 165 as sketch with colours, *Trigla volitans* L. —— Brasil, S. Parkinson.

NOTES: possibly because this fish was identified with the Linnaean species *Trigla volitans* it was not described. A number of other Brazilian fishes, although drawn, are not to be found in any manuscript, and for some reason were not described or the Brazilian animal manuscript has been lost.

196.(**2**:104) *Chelidonichthys kumu* (Lesson & Garnot, 1826) Triglidae

DRAWING: unfinished water-colour; *r.* [ink] 'S. Parkinson' [pencil, notes on coloration written on the drawing]; *v.* [pencil] '5. Trigla papilionacea/[ink] Opoorage'. 297 × 465.

MANUSCRIPT: Solander – (D. & W. 40a) P.A. f. 20 (22) as Trigla papilionacea, habitat Tolaga, Opoorage, two unmarked specimens preserved; f. 73 index to the specimens lists three preserved specimens. Dryander – Catalogue f. 165 as sketch with colours, *Trigla* —— New Zealand, S. Parkinson.

NOTES: Solander's name *Trigla papilionacea* was published by Cuvier (1829), although in the synonymy of *T. kumu*. Cuvier referred to this Parkinson drawing in Banks's library. This drawing was reproduced by Whitehead (1968) as Plate 35.

Both Parkinson's drawing and Solander's manuscript were referred to by Richardson (1843*a* & *b*) under the name *Trigla papilionacea*.

197.(**2**:105) *Bagre marinus* (Mitchill, 1815) Ariidae

DRAWING: finished water-colour; *r.* [ink] 'Silurus – Bagra-/Sydney Parkinson pinx^t^ 1769'; *v.* [pencil] 'The fins Gray/7 Silurus Bagre/[ink] Brasil'. 294 × 467.

MANUSCRIPT: Solander – not traced. Dryander – Catalogue f.167 as finished in colours, *Silurus bagre* L. —— Brasil, S. Parkinson.

NOTES: this appears to be another example of a Brazilian fish which, although drawn by one of the artists, was not described by Solander.

198.(**2**:106) *Sphyraena helleri* Jenkins, 1901 — Sphyraenidae

DRAWING: unfinished water-colour; *r.* [ink] 'S. Parkinson'; *v.* [pencil] 'N° 2 Esox sphyraenoides/[ink] Otahite'. 299 × 479.

MANUSCRIPT: Solander – (D. & W. 40c) P.A.O.P. f.60 (180) as Esox Sphyrenoides, f.290 index, specimen not preserved. Dryander – Catalogue f.171, one of two drawings of *Esox* ——, sketch with colours, Society Islands, S. Parkinson.

NOTES: Solander's name Esox sphyraenoides does not seem to have been taken up by later naturalists, although Cuvier (1829) cites a drawing in Banks's library made at Tahiti and labelled Esox sphyraenoides which he attributed to G. A. Forster. However, there are no drawings of *Sphyraena* in the Forster collection, nor is there a drawing labelled Esox sphyraenoides (Wheeler, 1981), so it must be assumed that Cuvier mistook this Parkinson drawing for one by Forster. This would be understandable as he worked only from copies of the drawings. However, the confusion was unfortunate in that he named the species *Sphyraena forsteri* under the false impression that it was a Forster specimen.

199.(**2**:107) *Platybelone argala* (Le Sueur, 1821) — Belonidae

DRAWING: unfinished water-colour; *r.* [pencil] 'Ihre Eawaou/Ichea Eawaou/Es. belone L./[ink] S. Parkinson'; *v.* [pencil] 'N° 65 Esox rostratus/[ink] Otahite'. 268 × 374.

MANUSCRIPT: Solander – (D. & W. 40c) P.A.O.P. f.26 (144), as Esox rostratus, three specimens preserved numbered A.60. Dryander – Catalogue f.171, one of two drawings of *Esox* —— sketch with colours, Society Islands, S. Parkinson.

NOTES: Solander's name Esox rostratus seems not to have been adopted by later authors, although the specific epithet has been independently employed for other species of garfish. The identification of this drawing with the Linnaean *Esox belone*, the only garfish to have been named at the time of the *Endeavour* voyage was probably by Broussonet.

200.(**2**:108) *Parexocoetus brachypterus* (Richardson, 1846) — Exocoetidae

DRAWING: unfinished water-colour; *r.* [pencil] 'Etèpa/[ink] S. Parkinson'; *v.* [pencil] 'The roundness of the back to be taken of to where it is mark'd/the blue to be ultramarine/N° 44 Exocoetus brachyopterus/[ink] Otahite'. 268 × 371.

MANUSCRIPT: Solander – (D. & W. 40c) P.A.O.P. f.13 (129) as Exocoetus brachyopterus, George Land ten specimens numbered A.35. Dryander – Catalogue f.173, sketch with colours *Exocoetus* —— Society Islands, S. Parkinson.

NOTES: the Solander name *Exocoetus brachypterus* was published by Richardson (1846) within his discussion of *Exocoetus monocirrhus*. He referred to both the Parkinson drawing and the Solander manuscript. Both manuscript and drawing were also referred to by Günther (1866), who also gave a measurement of the drawing. This drawing therefore has type status.

The manuscript and the drawing were also cited by Valenciennes (1847) in his description of *Exocoetus Solandri* and form the bases for the foundation of that species name. However, Valenciennes wrongly attributed the drawing to Forster (an error he made elsewhere); the Forster drawings include only one flyingfish and that drawn off the European coast (Wheeler, 1981).

This drawing was reproduced by Whitehead (1968) as Plate 9.

201.(**2**:109) *Cypselurus* (*Poecilocypselurus*) *poecilopterus* (Valenciennes in Cuvier & Valenciennes, 1847) Exocoetidae

DRAWING: unfinished water-colour; *r.* [pencil] 'E. volitans/Mararaa/[ink] S. Parkinson'; *v.* [pencil] 'lower part of the head & eye & fore part of the belly silver the Pinnae transparent/the spots on the P.P. black./N°. 4 Exocoetus alatus/[ink] Otahite'. 289 × 475.

MANUSCRIPT: Solander – (D. & W. 40c) P.A.O.P. f.33 (151) as Exocoetus alatus, two specimens numbered A.67. Dryander – Catalogue f.173, sketch with colours *Exocoetus volitans* —— Ocean, S. Parkinson.

NOTES: Solander's description, but not the drawing for certain, was cited by Valenciennes (1847) although he had a specimen collected on the *Astrolabe* expedition for his main description of *Exocoetus poecilopterus*.

202.(**2**:110) *Exocoetus volitans* Linnaeus, 1758 Exocoetidae

DRAWING: five pencil studies of flying fishes and fins; *r.* [pencil] 'E. evolans L/ [ink] S. Parkinson'; *v.* [pencil] 'Mem the back is of a blackish blue mix[t] with brown which turns paler/toward the side & goes gradually into a silver colour the fins all transparent/the pupil of the eye very dark blue the iride dark brown the top of the/back part of the head is very brown, the tail gray underneath upon where therein/lines is mark'd with fine strip of blue/N° 2. Exocoetus volitans'. 295 × 475.

MANUSCRIPT: Solander – (D. & W. 45) Pisces 2, f.213–215v, as *Exocoetus volitans* L., Habitat in Oceano Atlantico; (D. & W. 42) C.S.D., same data. Dryander – Catalogue f.173, sketch without colours, *Exocoetus volitans* L. —— Ocean, S. Parkinson.

NOTES: this drawing was referred to by Richardson (1846) as *Exocoetus volans* Solander, and he also cited the Solander manuscripts. Richardson's only other material was a Chinese fish collected by Sir Edward Belcher, and as he gave 'Seas of China and Polynesia' as the habitat of the species it seems that he wrongly assumed that this *Endeavour* specimen came from Polynesia. It seems from Solander's notes, however, that it was an Atlantic specimen which was described and drawn.

203.(**2**:111) *Eleutheronema tetradactylum* (Shaw, 1804) Polynemidae

DRAWING: unfinished water-colour; *r.* [ink] 'S. Parkinson'; *v.* [pencil] 'Polynemus quadrenarius/[ink] Endeavours river'. 268 × 372.

MANUSCRIPT: Solander – (D. & W. 40b) P.N.H. f.15 (97) as Polynemus quadernarius, habitat near Endeavour River Careening place 30 July 1770. Dryander – Catalogue f.173, no Parkinson drawing entered, two of the three Forster drawings listed are probably misattributed.

NOTES: Solander's name Polynemus quadernarius seems not to have been adopted by later writers.

204.(**2**:112) *Coris gaimardi* (Quoy & Gaimard, 1824) Labridae

DRAWING: unfinished water-colour; *r.* [pencilled coloration notes on figure], [ink] 'S. Parkinson'; *v.* [pencil] 'the strips on the head verditer & Gamboge the body is a purple black spotted especially toward the tail with/Ultramarine & towards the head w[t] small spots of Green – the border, stripes & spots on the tail blue/the body turns more purple towards the head/[pencil] 71/[ink] Ulhietea/[pencil, on mount of drawing] Julis gaimardi Frey Voy Uranie t54 f1'. 271 × 374.

MANUSCRIPT: Solander – not traced, was not named and probably never described. Dryander – Catalogue not traced.

NOTES: there is no evidence that this fish was named, and therefore it is impossible to relate it to either Dryander's catalogue or Solander's manuscript. The specimen was collected at Raiatea (16°50′ S., 151°24′ W.).

205.(**2**:113) *Saurida gracilis* (Quoy & Gaimard, 1824) Synodontidae

DRAWING: unfinished water-colour, pencil detail of jaw dentition; *r.* [pencil] 'Arāi [ink] S. Parkinson'; *v.* [pencil] 'The edges of the scales border'd w[t] brownish purple very dark where the spot is & on the back towards the tail./N[o] 22 Dentex nebulosus/[ink] Otahite'. 267 × 369.

MANUSCRIPT: Solander – (D. & W. 40c) P.A.O.P. f.6 (118) as Dentex nebulosus, serial number A.14 but not so labelled in Cagg 3; (f.285) index shows that two specimens were preserved. Dryander – Catalogue f.169 as Dentex ——, sketch with colours, Society Islands, S. Parkinson.

NOTES: the Solander name Dentex nebulosus was published by Valenciennes (1849) as *Saurida nebulosa*. Although Valenciennes had specimens from the Ile de France (Mauritius) he also referred to the Parkinson drawing which he had seen. Solander's genus Dentex was an independent proposal to *Dentex* of Cuvier (1815) in the family Sparidae.

206.(**2**:114) *Synodus variegatus* (Quoy & Gaimard, 1824) Synodontidae

DRAWING: unfinished water-colour; *r.* [pencil] 'Arai/[ink] S. Parkinson'; *v.* [pencil] 'N[o] 47. Dentex marmoreus/[ink] Otahite'. 267 × 371.

Fig. 13 *Saurida gracilis* (Quoy & Gaimard, 1824). Drawing by Parkinson of one of two specimens caught in Tahiti. (Catalogue number 205.)

MANUSCRIPT: Solander – (D. & W. 40c) P.A.O.P. f.83 (203) as Dentex marmoreus, one specimen not numbered, another numbered A.135; (f.290) index lists two specimens. Dryander – Catalogue f.169 as Dentex ——, sketch with colours, Society Islands, S. Parkinson.

NOTES: Valenciennes (1849) referred to a copy of this drawing and also had a copy of the manuscript description of Dentex marmoreus. It is interesting that at that date he correctly cited the history of the drawing as in the library of Banks, 'aujourd'hui déposée au British Museum', but that the copies had been made by permission of Robert Brown.

207.(**2**:115*a*) *Gomphosus varius* Lacepède, 1802 Labridae

DRAWING: unfinished water-colour; *r.* [pencil] 'Paootiroa/Páoùdúróá/[ink] S. Parkinson'; *v.* [pencil] 'N° 7 Nasutus virescens/[ink] Otahite'. 239 × 340.

MANUSCRIPT: Solander – (D. & W. 40c) f.49 (169) as Nasutus virescens, one specimen, numbered A.95 in Cagg 4; (f.289), same data. Dryander – Catalogue f.153 as Nasutus ——, sketch with colours, Fr. Isl. (= Friendly Islands), S. Parkinson.

NOTES: Solander's appropriate genus name, Nasutus, for the bird-wrasses was never formally published and they were unnamed until Lacepède proposed *Gomphosus* in 1802. Two specimens still exist in the British Museum (Natural History) collections, S.L. 105 and 145 mm, and were listed by Günther (1862) under *Gomphosus varius* with a note of an old label reading Labrus nasutus on it (the writing on this label is still readable). There seems to be no doubt that those specimens are *Endeavour* specimens and represent this fish and the succeeding one.

Dryander's entry of Friendly Islands as the locality for this drawing is erroneous; these islands were not visited by the *Endeavour*.

Valenciennes (1840) in his discussion of *Gomphosus fuscus* refers to this drawing in Banks's library, noting that it was captioned nasutus.

The drawing represents a terminal phase male.

208.(**2**:115*b*) *Gomphosus varius* Lacepède, 1802 Labridae

DRAWING: unfinished water-colour; *r.* [pencil] 'Epàoū outouròā/[ink] S. Parkinson'; *v.* [pencil] 'narrow border of white upon the P.D. & ye P.A. the cheeks silvery/the back especialy towards the tail darker & the colour on the edges./The scales their considerably broader. the strip on the tail a dilute red or white w^{t} red cast/N° 23 Nasutus purpureus/nasutus/[ink] Otahite'. 203 × 283.

MANUSCRIPT: Solander – (D. & W. 40c) P.A.O.P. f.3 (115) as Nasutus purpurascens, serial number A.11, George Land; f.48 (166) one specimen serial number A.94; (f.285) index confirms that two specimens were preserved. Dryander – Catalogue f.153 as Nasutus ——, sketch with colours, Fr. Isl., S. Parkinson.

NOTES: see notes in number 207 above.

This drawing represents a female fish turning into a male.

209.(**2**:116) *Lutjanus fulvus* (Bloch & Schneider, 1801) Lutjanidae

DRAWING: unfinished water-colour; *r.* [ink] 'S. Parkinson/[pencil] Ettoou'; *v.* [pencil] '70 [ink] Ulhietea'. 270 × 371.

MANUSCRIPT: Solander – not traced. Dryander – Catalogue not traced.

NOTES: as this fish was not named it is impossible to trace it in either the manuscript or the catalogue of drawings.

210.(**2**:117) *Centropyge bispinosus* (Günther, 1860) Pomacanthidae

DRAWING: unfinished water-colour; *r.* [pencil] 'Hoomurea/Athòdi tui tui/[ink] S. Parkinson'; *v.* [ink] 'Otahite'. 235 × 298.

MANUSCRIPT: Solander – not traced. Dryander – Catalogue not traced.

NOTES: as the fish in this drawing was not identified it is impossible to trace it in either the manuscript or the catalogue of drawings. It is perhaps worth noting here that the arrangement of the fish drawings followed Linnaeus's twelfth edition of the *Systema Naturae*, but that the six last drawings (the two unrecognized genera, Dentex and Nasutus, and the two unidentified fishes) could not be accommodated in the system and were merely grouped at the end of the volume.

211.(**3**:1*a*) *Leucophaea maderae* (Fabricius, 1781) Oxyhaloidae

DRAWING: water-colour, dorsal and ventral views of insect, and egg case by A. Buchan; *r.* [pencil] 'Blatta maderae'; *v.* [none]. 133 × 232.

MANUSCRIPT: Solander – (D. & W. 45) S.C. Hemiptera f. 2 as Blatta domestica 'in Madeira culinis'. Dryander – Catalogue f. 191 as finished in colours, Blatta Madera Fabr. —— Madeira, A. Buchan.

NOTES: Fabricius (1781) based his name on a specimen examined in 'Mus. Dom Banks' which, because of the locality and of the identification given to this drawing by Dryander must have been the specimen drawn by Buchan (or one collected at the same time). The drawing therefore has some standing as type material.

It is of interest to note that the Dryander Catalogue lists two drawings by Buchan of *Blatta germanica* Linnaeus, 1767 with the locality of 'in nave'. Neither of these is now in the collection of *Endeavour* drawings.

212.(**3**:1*b*) Isoptera (family, genus and species indet) Order Isoptera

DRAWING: water-colour, dorsal view, by A. Buchan; *r.* [pencil] 'Termes fatale/ Winged white ant/Termes Fatale'; *v.* [ink] 'Rio Janeiro'. 233 × 257.

MANUSCRIPT: Solander – no description found. Dryander – Catalogue f. 203 as finished in colour Termes fatalis L. —— Brasil, A. Buchan.

NOTES: due to the difficulty in identifying termite specimens it is not possible to give this drawing any identification. *Termes fatale* was a name given by Linnaeus (1758), and was the only species in the genus recognized by Fabricius (1775). Fabricius quoted the Banks Collection in his description but in his opening paragraph referred to the insect as 'Habitat in Indiae . . .', clearly not referring to this drawing. Nevertheless the pencil annotations are believed to be in Fabricius's hand.

213.(**3**.2*a*) *Sceliphron coementarium* (Drury, 1770) Sphecidae

DRAWING: water-colour of a single mud-dauber wasp, lateral view by A. Buchan; *r.* [no annotations]; *v.* [ink] 'Madera'. 182 × 265.

MANUSCRIPT: Solander – no description found. Dryander – Catalogue f. 205 as finished in colour, Sphex —— Madeira, A. Buchan.

NOTES: Drury's (1770) account of *Sphex coementarium* was based on specimens from 'Antigua, St. Christopher's, and Jamaica'. At the time of this drawing this

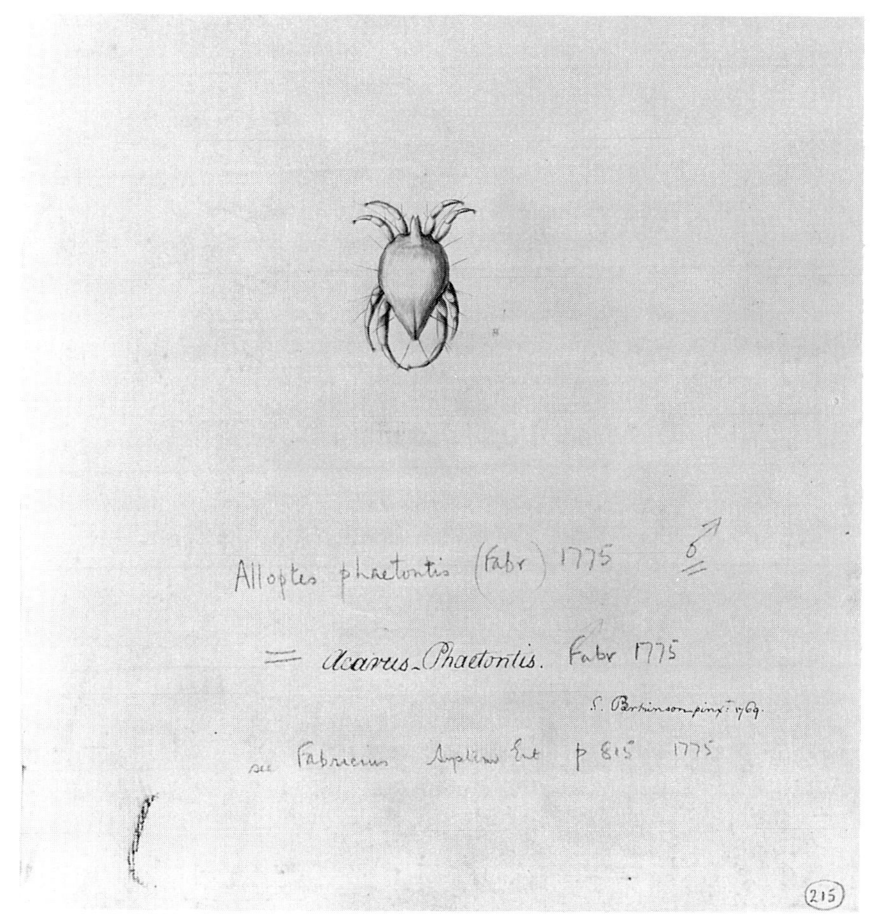

Fig. 14 *Laminalloptes phaetontis* (Fabricius, 1775). Parkinson's drawing of a feather mite taken from a red-tailed tropic bird shot near Tahiti in March 1769. The drawing was used by Fabricius for his description. (Catalogue number 215.)

mud-dauber wasp was unknown to science; the species was not recorded from Madeira until 1825 (M. Day, personal communication 1979).

214.(**3**:2*b*) *Sceliphron coementarium* (Drury, 1770) Sphecidae

DRAWING: water-colour, two dorsal, one lateral and one oblique views by A. Buchan; *r.* [no annotations]; *v.* [ink] 'Madera'. 183 × 264.

MANUSCRIPT: Solander – no description found. Dryander – Catalogue f.205 as finished in colour, Sphex —— Madeira, A. Buchan.

NOTES: see no.213 in this catalogue.

215.(**3**:3) *Laminalloptes phaetontis* (Fabricius, 1775) Proctophyllodidae

DRAWING: pencil; *r.* [pencil] 'Alloptes phaetontis (Fabr) 1775 ♂ = [ink] Acarus

Phaetontis. [pencil] Fabr 1775/[ink] S. Parkinson pinxt. 1769/[pencil] see Fabricius System Ent p 815 1775'; *v.* [ink] 'March 21 1769/Lat 25.21′ Long. 139 W.'. 288 × 234.

MANUSCRIPT: Solander – (D. & W. 45) S.C. Diptera & Aptera f. 101 as Acarus phaetontis on Phaetontis in Oceano Australe; (D. & W. 42) C.S.D. f.291, as above, 'Habitat copiose in Phaetonte erubescente'. Dryander – Catalogue f. 213 as finished in colours, Acarus Phaëtontis Mss —— Ocean, S. Parkinson.

NOTES: Fabricius (1775) appears to have relied entirely on the Banks material for his description, writing 'Habitat in Phaetonte erubescente Oceani australis Fig. pict *Mus Banks.*' It is uncertain whether this should be read to imply that a specimen was available in the Banks collection, or whether it was the drawing (and Solander's manuscript description) which was available in Banks's museum, as well as the specimen, the latter being the more probable. In any case this drawing has type standing.

The source of Fabricius's information, namely the Solander manuscript description and the Parkinson drawing, was unknown to Atyeo & Peterson (1967), who questioned the identity of the feather mite named by Fabricius and were doubtful concerning the identity of the host bird.

Several of the annotations on the recto quoted above are twentieth-century additions, only the words 'Acarus Phaetontis' and Parkinson's signature are contemporary with the drawing. In fact, Solander's *Phaeton erubescens* is referable to *Phaethon rubricauda melanorhynchos* Gmelin, 1789 (and his material provided the basis for Gmelin's name). Atyeo & Peterson (1967), following earlier workers, considered that the host was *Phaeton lepturus fulvus*; this appears to be incorrect.

216.(**3**:4*a*) *Argiope bruennichi* (Scopoli, 1772) Araneidae

DRAWING: finished water-colour, dorsal and ventral views by A. Buchan; *r.* [pencil] 'Argyope bruennichii (Scopoli) 1772/= Aranea fasciata Fabr 1775/ Fabricius Syst Ent p.433 1775'; *v.* [ink] 'Madera'. 146 × 236.

MANUSCRIPT: Solander – none. Dryander – Catalogue f.215 as finished in colour, Aranea —— Madeira, A. Buchan.

NOTES: this drawing, and possibly a specimen in Banks's collection, were used by Fabricius (1775) as the basis of his *Aranea fasciata*. The drawing therefore has some standing as type material.

All the annotations on the recto are recent.

217.(**3**:4*b*) *Nephilgenys cruentata* (Fabricius, 1775) Araneidae

DRAWING: finished water-colour; *r.* [pencil] 'Nephilengys cruentata Fabr. 1775/ Fabricius Syst. Ent. p439 1775./[ink] Sydney Parkinson pinx[t] ad vivum'; *v.* [pencil] 'Betranea/[ink] Rio de Janeiro'. 229 × 267.

MANUSCRIPT: Solander – none. Dryander – Catalogue f.215 as finished in colour, Aranea —— Madeira, S. Parkinson.

Fig. 15 *Sesarma* sp. Drawn by Buchan at Funchal, Madeira. This drawing may have been consulted by Fabricius (1787) when he described the species *Cancer quadratus* from Banks's material. (Catalogue number 218.)

NOTES: Fabricius (1775) species *Aranea cruentata* is based solely on material in Banks's collection; he wrote 'Habitat ad Rio Janeiro Brasiliae. *Mus. Dom. Banks.*'. Whether he saw a specimen in Banks's collection or just the drawing is not known but the drawing has type status. However, Fabricius apparently did not label this drawing, nor the others he described which are discussed above.

It is presumed that Dryander's entry of 'Madeira' for this drawing in his Catalogue was a *lapsus calami* for Brazil, as the drawing is so labelled and Fabricius clearly reported the species from that country.

218.(**3**:5) *Sesarma* sp. Grapsidae

DRAWING: finished water-colour by A. Buchan; *r.* [pencil] 'Cancer quadratus/ – mutus L./Buchan'; *v.* [ink] 'Madera'. 170 × 265.

MANUSCRIPT: Solander – (D. & W. 45) S.C. Diptera & Aptera f.113, as Cancer quadratus, Funchal, Madeira. Dryander – Catalogue f.215 as finished in colour, *Cancer mutus* L. —— Madeira, A. Buchan.

NOTES: the name *Cancer quadratus* was published by Fabricius (1787) based on a specimen said to have 'Habitat in Jamaica Mus. Dom. Banks'. It is not possible to be certain whether Jamaica was written in error for Madeira; if so, Fabricius's type material was possibly the *Endeavour* specimen and this drawing has type status. At least the specific epithet must have been derived from Solander's manuscript. However, Fabricius later (1798) published *Cancer quadratus* independently on a specimen from 'India orientali, Dom. Daldorff' and this clearly owes nothing to the *Endeavour* material.

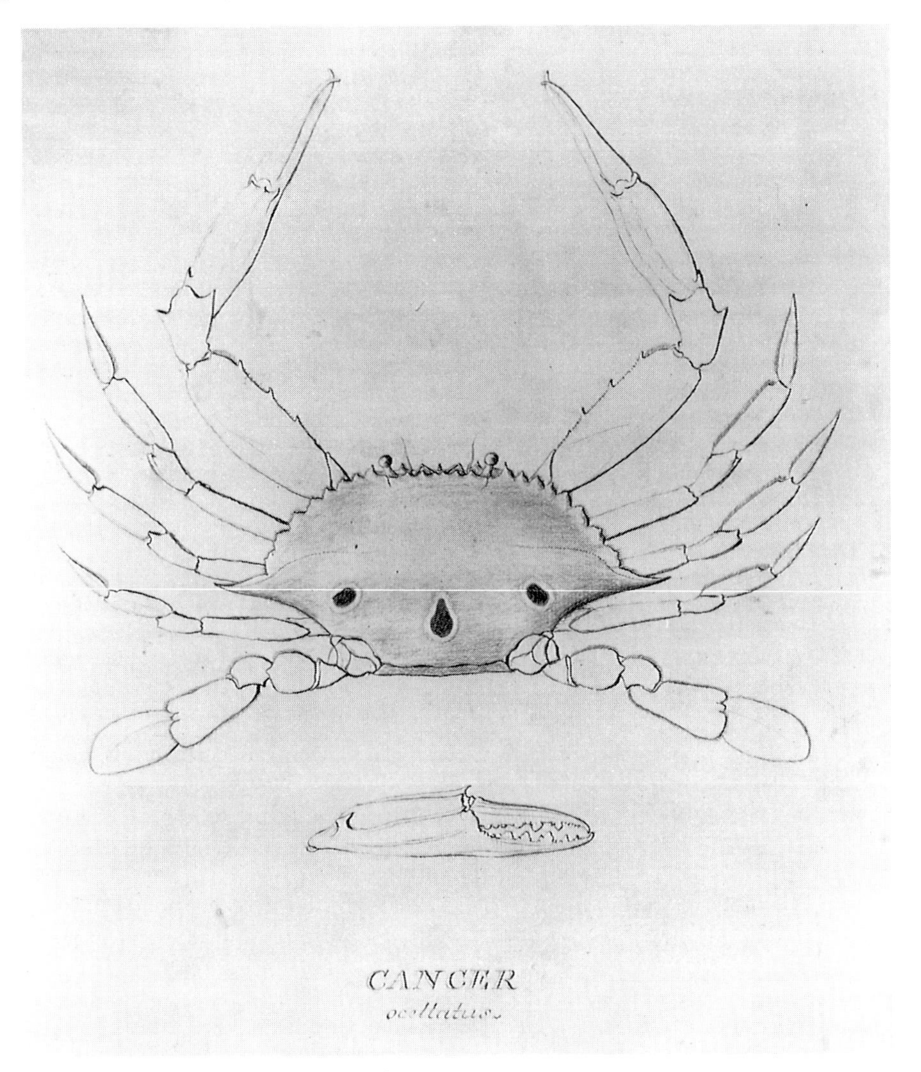

Fig. 16 *Portunus sanguinolentus* (Herbst, 1783). Drawing by Spöring of a swimming crab captured off the Australian coast. (Catalogue number 219.)

219.(**3**:6) *Portunus sanguinolentus* (Herbst, 1783) Portunidae

DRAWING: finished pencil drawing by H. D. Spöring; *r.* [pencil] 'CANCER/ ocellatus'; *v.* [none]. 362 × 344.

MANUSCRIPT: Solander – (D. & W. 45) S.C. Diptera & Aptera f.116 as Cancer ocellatus, New Holland. Dryander – Catalogue f.215 as finished without colour Cancer ocellatus Mss —— N.C. (= Nova Cambria), Spöring.

NOTES: the name *Cancer ocellatus* was published by Herbst (1799) but from another specimen collected at Long Island, New York; it appears to have been a quite independent proposal to that of Solander.

220.(**3**:7) *Portunus pelagicus* (Linnaeus, 1758) Portunidae

DRAWING: finished pencil drawing by H. D. Spöring; *r.* [pencil] 'CANCER/ pelagicus'; *v.* [none]. 376 × 372.

MANUSCRIPT: Solander – (D. & W. 45) S.C. Diptera & Aptera f.118 as *Cancer pelagicus*, New Holland. Dryander – Catalogue f.215 as finished without colour, *Cancer pelagicus* L —— N.C. (= Nova Cambria), Spöring.

NOTES: Solander had evidently associated the specimen from which this drawing was made with Linnaeus's species, *Cancer pelagicus*. It is drawn on a large sheet of paper which was folded in binding and which had become torn along the fold.

221.(**3**:8) *Polybius henslowii* Leach, 1820 Portunidae

DRAWING: finished water-colour by A. Buchan; *r.* [ink] 'CANCER Depurator/ Sept. 4. 1768'; *v.* [ink] 'off the coast of Spain. Sept$^{r.}$ 4$^{th.}$ 1768'. 235 × 291.

MANUSCRIPT: Solander – (D. & W. 45) S.C. Diptera & Aptera f. 125 as Cancer depurator, Atlantic Ocean; (D. & W. 42) C.S.D. f.327 as C. Depurator 'Hab. in Oceano Atlantico, ubi non procul a Capite Finnisterra copiose in superficium aqua visus & facile retibus (vulgo cast-nets) captus. Sept. 3, 1768'. Dryander – Catalogue f.215 as finished in colour, *Cancer depurator* L. —— Ocean, A. Buchan.

NOTES: Solander misidentified this, as then undescribed, swimming crab with the Linnaean species *Cancer depurator* (= *Liocarcinus depurator*, see Ingle (1980)). The note quoted from the manuscript (C.S.D.) is one of the few occasions when observations of the behaviour of the animal, and its means of capture, were recorded. It also shows that the artist was working on the material a day after the animal's capture (although it may merely have been signed on completion then). This manuscript account also records the drawing by number (Fig. Pict. N. 11) confirming that the drawings were numbered serially within each region visited, and suggests that there was a master list of drawings by all the artists.

222.(**3**:9) *Munida gregaria* (Fabricius, 1775) Galatheidae

DRAWING: finished water-colour, upper-dorsal view, lower-lateral view; *r.* [ink] 'Cancer gregarius./Sydney Parkinson pinx^{t} 1769'; *v.* [ink] 'Jan$^{y.}$ 2nd 1769/Lat 37.30′.' 292 × 235.

MANUSCRIPT: Solander – (D. & W. 45) S.C. Diptera & Aptera f.153 as Cancer gregarius, off Patagonia; (D. & W. 42) C.S.D. f.321 as above, Habitat 'Oceano America australis ubi prope Patagoniam gregatium supra aquam natant, & colore suo rubro sape mare grasi cruentum reddunt (Jan. 2 1768). Lat 45°31′S, Long (Lond.) 61°29′W'. Dryander – Catalogue f.215 as finished in colour, Cancer Pagurus [inserted] gregarius Fabricius —— Ocean, S. Parkinson.

NOTES: Fabricius (1775) based his species *Pagurus gregarius* solely on material in Banks's collection. Whether he examined the specimens collected or merely made use of Parkinson's drawing and Solander's description is not known, but similarities in wording in his account suggest that he closely followed Solander's notes. This drawing therefore has some standing as type material.

This drawing was reproduced by Wheeler (1983) as Plate 189a.

223.(**3**:10) Megalopal stage of a paguroidean crab. Superfamily Paguridea

DRAWING: finished pencil and wash, left – dorsal view, centre – natural size, right – ventral view; *r.* [ink] 'Cancer amplectans/Sydney Parkinson pinx[t] ad vivum 1768'; *v.* [ink] 'Oct[r]. 30 1768 Lat. [blank] S.'. 247 × 293.

MANUSCRIPT: Solander – (D. & W. 45) S.C. Diptera & Aptera f.142 as Cancer amplectans, Atlantic Ocean off Brazil; (D. & W. 42) C.S.D. 307, 309, same data 'Noctu fulget'. Dryander – Catalogue f.215, as finished without colours, Cancer Pagurus amplectens Fabricius —— Ocean, S. Parkinson.

NOTES: the name *Pagurus amplectens* published by Fabricius (1775) was accompanied by the statement 'Habitat in Oceano atlantico, Brasiliam alluente, noctu fulgens'. Although Fabricius made no reference to Banks's collection for this species the coincidence between the use of the Solander name and the locality data shows that it was based on *Endeavour* material. This drawing therefore has type standing.

The megalop illustrated shows a combination of paguroidean and brachyuran features and its identification is questionable (R. W. Ingle – personal communication).

224.(**3**:11) Megalopal stage of a brachyrhynchan crab. Section Brachyrhyncha

DRAWING: finished pencil and wash, dorsal view; *r.* [none]; *v.* [pencil] 'Cancer cyanopthalmus/[ink] New Holland April 28, 1770'. 292 × 235.

MANUSCRIPT: Solander – none. Dryander – Catalogue f.217 as finished in colour, Cancer —— N.C. (= New Caledonia), S. Parkinson.

NOTES: there are no Solander notes relating to a specimen with the name of Cancer cyanopthalmus, and he may never have described it. The crustacean named *Portunus cyanophthalmus* by Peron (1807) was an independent use of the specific epithet based on a specimen, also from off New Holland 'caught 4 April 1802 – la mer parut coverte d'une charmante espèce de Portune remarquable . . . par la belle couleur bleue de ses deux yeux'.

225.(**3**:12) *Hippolyte caerulescens* (Fabricius, 1775) Hippolytidae

DRAWING: left – finished pencil, dorsal view enlarged, right – finished watercolour, natural size; *r.* [ink] 'Cancer caerulescens./Sydney Parkinson pinx[t] ad vivum 1768'; *v.* [ink] 'Oct[r]. 7. 1768'. 235 × 284.

MANUSCRIPT: Solander – (D. & W. 45) S.C. Diptera & Aptera f.136 as Cancer caerulescens, tropical Atlantic Ocean; (D. & W. 42) C.S.D. f.301, 303, Habitat in Pelago ubi intra Tropicos Oceani Atlantici frequens. Dryander – Catalogue f.217 finished in colour, Cancer Astacus caerulescens Fabr. —— Oc., S. Parkinson.

NOTES: Fabricius (1775) based his name *Astacus caerulescens* on material in Banks's collection, citing the locality as 'Habitat in Pelago inter Tropicos frequens'. From the similarity in wording it might be assumed that he was quoting from Solander's manuscript, and this drawing has to be accorded some type standing.

226.(**3**:13) Indeterminate euphausiacean or mysidacean. Order Euphausiacea

DRAWING: finished pencil, above – lateral view, below – lateral natural size; *r.* [ink] 'Cancer fulgens./Sydney Parkinson pinx^t 1768'; *v.* [ink] 'Octr. 30. 1768/Lat S'. 239 × 290.

MANUSCRIPT: Solander – (D. & W. 45) S.C. Diptera & Aptera f.145 as Cancer fulgens, off Brasil at surface; (D. & W. 42) C.S.D. f.311 as C. fulgens 'Habitat in Pelago Brasiliam alluente. Noctu fulges'. Dryander – Catalogue f.217, as finished without colour, Cancer Astacus fulgens Fabr. —— Ocean, S. Parkinson.

NOTES: the Solander name *Cancer fulgens* was published by Fabricius (1775), based entirely on material in Banks's collection. Fabricius's ascription 'Habitat in Oceano Brasiliam alluente, noctu fulgens. Mus. Banks.' so nearly coincides with Solander's manuscript that it must have been derived from it. This drawing therefore has some type status.

This illustration was reproduced by Macartney (1810, figs. 1–2) and information on the luminous ability of *Cancer fulgens* communicated by Sir Joseph Banks in Macartney's account of luminous animals.

227.(**3**:14) *Scina crassicornis* (Fabricius, 1775) Scinidae

DRAWING: finished pencil, left – natural size, right – enlarged; *r.* [ink] 'Cancer crassicornis./Sydney Parkinson pinx^t 1768'; *v.* [ink] 'Octr. 30 [deleted] 1768/Lat S.'. 292 × 236.

MANUSCRIPT: Solander – (D. & W. 45) S.C. Diptera & Aptera f.150 as Cancer crassicornis, off Brasil at the surface; (D. & W. 42) C.S.D. f.319, same details, 'Habitat intra tropicos in Pelago Brasiliano'. Dryander – Catalogue f.217 as finished without colour, Cancer Astacus crassicornis Fabricius —— Ocean, S. Parkinson.

NOTES: Fabricius (1775) published the name *Astacus crassicornis* attributing to it the locality 'Habitat in Oceano americano' and the source 'Mus. Banks'. There is no doubt that he referred to Solander's manuscript description which can be presumed to have been derived from the specimen Parkinson drew, and this drawing therefore has type standing.

Parkinson's drawing was referred to by Stebbing (1888), while it was under the care of Dr Albert Günther, then Keeper of the Department of Zoology, and he wrote '*Astacus crassicornis* of Fabricius is the earliest described species of the genus . . . while it is beyond all question that Sydney Parkinson's figure of Cancer crassicornis is the earliest known representation of any species of that genus'. Stebbing, however, referred only to Fabricius for the text on the species and appears not to have referred to Solander's more detailed description in his manuscript.

This drawing was reproduced by Wheeler (1983) at Plate 189c.

228.(**3**:15) *Lysierichthus vitreus* (Fabricius, 1775) Lysiosquillidae

DRAWING: finished pencil, left – enlarged dorsal view, centre – natural size, lateral view, right – enlarged ventral view; *r.* [ink] 'Cancer vitreus/Sydney Parkinson pinx^{t} 1768'; *v.* [ink] 'Coast of Brasil'. 239 × 294.

MANUSCRIPT: Solander – (D. & W. 45) S.C. Diptera & Aptera f.165 as Cancer vitreus, off Brasil, at surface; (D. & W. 42) C.S.D. f.337 as C. vitreus, Habitat in Pelago Oceani Atlantici non procul a Brasilia. Dryander – Catalogue f.217, as finished without colour, Cancer Squilla vitrea Fabricius —— Brasil, S. Parkinson.

NOTES: Fabricius (1775) based his species *Astacus vitreus* on material in Banks's collection, most probably the drawing and manuscript account by Solander. This drawing therefore has type status. *Lysierichthus* is a larval generic name within the stomatopod crustaceans. It is possibly the larval stage of *Lysiosquilla scabricauda* (Lamarck, 1818) according to Gurney (1946) and Manning (1969).
This drawing was reproduced by Wheeler (1983) at Plate 189b.

229.(**3**:16) *Lysierichthus vitreus* (Fabricius, 1775) Lysiosquillidae
Diodon hystrix Linnaeus, 1758 Diodontidae
Conchoderma virgatum var. *hunteri* Darwin, 1851 Lepadidae

DRAWING: pencil sketches, unfinished water-colour of *Diodon* (lateral view); dorsal and ventral views of *Lysierichthus*; four lateral views of *Conchoderma*; *r.* [ink] 'C. vitreus/Diodon Erinaceus/Lepas pelluscens'; *v.* [none]. 286 × 235.

DRAWING: see Notes.

NOTES: these are clearly preliminary studies for the finished drawings of the three animals. For details of these finished drawings see entries 73 (*Diodon*), 228 (*Lysierichthus*), and 290 (*Conchoderma*) in this catalogue.

230.(**3**:17) *Caligus corpyphaenae* Steenstrup & Lütken, 1861 Caligidae

DRAWING: finished pencil and wash; left (Fig. 1) – ventral view, centre (Fig. 3) – natural size, right (Fig. 2) – dorsal view; *r.* [ink] 'Monoculus piscinus/Sydney Parkinson pinx^{t} 1768'; *v.* [ink] 'Octr 1768/L. N'. 239 × 296.

MANUSCRIPT: Solander – (D. & W. 45) S.C. Diptera & Aptera f.182 as Monoculus piscinus, on Scomber pelamidi, Atlantic Ocean; (D. & W. 42) C.S.D. f.347–349 as above. Dryander – Catalogue f.219 as finished without colour, Monoculus piscinus L. —— Ocean, S. Parkinson.

NOTES: *Monoculus piscinus*, the name used by Solander, was a Linnaean name dating from 1761; it is generally considered to be a composite taxon. This drawing was made from a specimen taken from a scombroid fish, presumably the skipjack tuna, *Katsuwonus pelamis* (Linnaeus, 1758) – see number 187 in this catalogue – from which two internal parasites were also described.

231.(**3**:18*a*) Unidentified hyperioid amphipod Phronimidae

DRAWING: finished pencil; seven views, numbered Fig. 1 to Fig. 7, of different aspects or parts of the amphipod, various appendages lettered; *r.* [ink] 'Onidium gibbosum/P/T.15 P. Sept. 7. 1768'; *v.* [none]. 122 × 234.

MANUSCRIPT: Solander – (D. & W. 45) S.C. Diptera & Aptera f.198 as Onidium gibbosum, near Portugal, Atlantic Ocean, inside *Dagysas*; (D. & W. 42) C.S.D. f.359, same data. Dryander – Catalogue f.219 as finished without colour, Onidium gibbosum mss —— Ocean, S. Parkinson.

NOTES: Fabricius's (1775) description of *Oniscus gibbosus* established this taxon on the sole basis of the Parkinson drawing, for his description includes the statement 'Habitat in Oceano Lusitanico. Fig.pict. in Mus. Bankiano'. This appears to suggest that he made no use of Solander's manuscript description, a surprising omission if true. As Stebbing (1888) points out this *Endeavour* material was the earliest to be figured (and described) of any of this group, i.e. in what he called the family Typhidae. It also appears to be the first recorded observation of the possibly symbiotic relationship between an hyperiid amphipod and a salp (Solander's *Dagysas*).

232.(**3**:18*b*) Unidentifiable hyperiid amphipod Hyperiidae

DRAWING: finished pencil drawing; Fig. 1, left – dorsal view, Fig. 2, right – natural size lateral view; *r.* [ink] 'Onidium oblongatum/P/T.16 P Sept. 7. 1768'; *v.* [no annotation; two sketches of an amphipod]. 119 × 235.

MANUSCRIPT: Solander – (D. & W. 45) S.C. Diptera & Aptera f.202, as Onidium oblongatum, Atlantic Ocean inside *Dagysas*; (D. & W. 42) C.S.D. f.361 'Habitat in Oceano Atlantico, communis cum *Onidio gibboso* inter Dagysas, quas permeat vulneratque (Sept 6, 1768)'. Dryander – Catalogue f.219 as finished without colour, Onidium oblongatum Mss —— Ocean, S. Parkinson.

NOTES: Stebbing (1888) refers to this drawing, having examined it (as he did the other amphipod drawings) but beyond referring to it as one of the Hyperina wrote that it was 'without sufficient enlargement to show clearly the position of the species . . .'. This, and the other drawings of hyperiid amphipods, were some of the earliest made by Parkinson on the voyage, and show the meticulous detail of his work with microscopic subjects.

Solander's name Onidium oblongatum appears not to have been employed by later naturalists.

233.(**3**:18*c*) *Hyperia medusarum* (O. F. Muller, 1776) Hyperiidae

DRAWING: finished pencil and water-colour; Fig. 1, left-side view, Fig. 2, right-oblique front view; *r.* [ink] 'Onidium quadricorne./[pencil] Sydney Parkinson pinx[t] ad vivum 1768/[ink] T.2.P.2. August.28. 1768'; *v.* [none]. 122 × 235.

MANUSCRIPT: Solander – (D. & W. 42) C.S.D. f.363 as Onidium quadricorne 'Habitat in mari Atlantico, inter Hiberniam & Galliam et prope litore Gallicia'. Dryander – Catalogue f.219 as finished in colour, Onidium quadricorne Mss —— Ocean, S. Parkinson.

NOTES: Fabricius (1775) published the Solander name as *Oniscus quadricornis* 'Habitat in mari Atlantico. Mus. Banks'; this drawing therefore has standing as type material. Stebbing (1888) in an appendix to the bibliography in his work on the Amphipoda of the *Challenger* expedition, lists all the relevant Parkinson drawings and discusses their taxonomic significance. He therefore recognized that *Onidium quadricorne* was a synonym of O. F. Muller's *Hyperia medusarum* (which name is given here) but misattributed the former species to Fabricius (1781) whereas as it was published in 1775 it has priority over Muller's name.

234.(**3**:19) *Cystosoma spinosum* (Fabricius, 1775) Cystosomatidae

DRAWING: finished pencil, left – ventral view, centre – dorsal view, right – lateral view, all enlarged and with appendages lettered; *r.* [ink] 'Onidium spinosum/ Sydney Parkinson pinx^{t} 1768'; *v.* [unfinished pencil sketches of same species/'Octr. 7th. 1768/Lat. N.'. 270 × 374.

MANUSCRIPT: Solander – (D. & W. 45) S.C. Diptera & Aptera f.206 as Onidium spinosum, Atlantic Ocean; (D. & W. 42) C.S.D. f.365, same data Latitude Sept. IX.43. Dryander – Catalogue f.219 as finished without colour, Onidium spinosum mss —— Ocean, S. Parkinson.

NOTES: Fabricius (1775) described *Oniscus spinosus* as 'Habitat in Oceano Atlantico. Mus. Dom. Banks'; this drawing therefore has type standing. This figure was examined and referred to by Stebbing (1888).

235.(**3**:20) *Cystosoma spinosum* (Fabricius, 1775) Cystosomatidae

DRAWING: unfinished pencil, left – ventral view, centre – dorsal view, right – lateral view, all enlarged and with appendages lettered; *r.* [ink] 'Onidium spinosum'; *v.* [none]. 236 × 290.

MANUSCRIPT: see number 234 in this catalogue.

NOTES: this must have been a preliminary sketch for number 234 in this catalogue.

236.(**3**:21*a*) *Sapphirina* sp. Sapphirinidae

DRAWING: finished pencil and wash, left, Fig. I – dorsal view, centre, F. III – dorsal view natural size, right, Fig. II – ventral view, all with appendages lettered; *r.* [ink] 'Carcinium-opalinum/Sydney Parkinson pinx^{t} ad vivum Sepr 5th 1768/ [unclear] Sept 5 1768'; *v.* [none]. 125 × 230.

MANUSCRIPT: Solander – (D. & W. 45) S.C. Diptera & Aptera f.192 as Carcinium opalinum, Atlantic Ocean, near France; (D. & W. 42) C.S.D. f.353 as above, September 4, 1768. Dryander – Catalogue f.221 as finished without colour, Carcinium copallinium (*sic*) mss —— Ocean, S. Parkinson.

NOTES: the name *Carcinum opalinum* was published in Hawkesworth's (1773) account of the voyage of the *Endeavour*, attributed by C. D. Sherborn (*Index Animalium, 1880–1850*) to Banks and Solander, although there is no evidence to suggest that Banks was involved in the analysis of the animal other than through

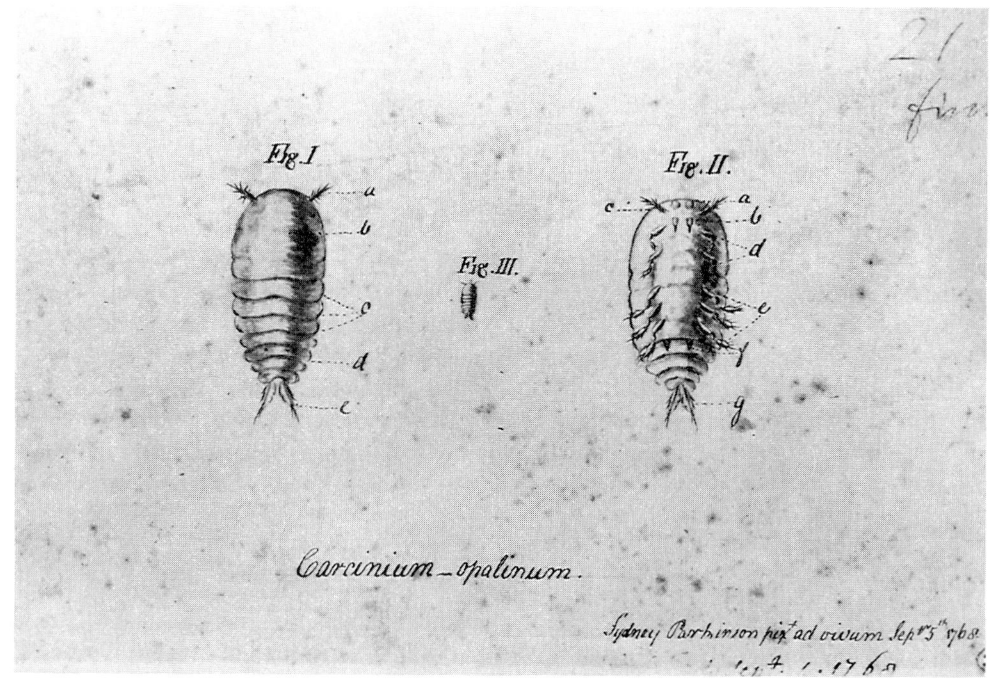

Fig. 17 *Sapphirina* sp. A detailed drawing by Parkinson of a copepod, the brilliant coloration of which was noted by Banks and Solander in Hawkesworth's (1773) account of the voyage. (Catalogue number 236.)

patronage. Hawkesworth wrote under September 1768 'Another animal of a new genus they also discovered, which shone in the water with colours still more beautiful and vivid, and which indeed exceeded in variety and brightness any thing that we had ever seen: the colouring and splendour of these animals were equal to those of an Opal, and from their resemblance to that gem, the genus was called *Carcinium Opalinum*'.

The drawing is not identifiable to species.

237.(**3**:21*b*) *Sapphirina* sp. Sapphirinidae

DRAWING: finished pencil and wash, left, Fig. 2 – ventral view of ovigerous female, centre, Fig. 3 – dorsal view natural size, right, Fig. 1 – dorsal view, both drawings 1 and 2 with appendages lettered; *r*. [ink] 'Carcinium macrourum/Sydney Parkinson pinx^{t} ad vivum Septr 7th 1768/Sept. 6. 1768.'; *v*. [none]. 125 × 231.

MANUSCRIPT: Solander – (D. & W. 45) S.C. Diptera & Aptera f.194 as Carcinium macrouram, Atlantic Ocean near Spain; (D. & W. 42) C.S.D. f.355, as above, 'Habitat in mari Atlantico Galliciam alluente Sept.4. 1768'. Dryander – Catalogue f.221 as finished without colour, Carcinium macrourum mss —— Ocean, S. Parkinson.

NOTES: the Solander name Carcinium macrourum does not appear to have been taken up by later naturalists.

238.(**3**:21*c*) *Idotea* sp. Idoteidae

DRAWING: finished water-colour by A. Buchan; above – dorsal view, below – ventral view; *r.* [ink] 'ONISCUS chelipes/Sept.2. 1768'; *v.* [none]. 113 × 230.

MANUSCRIPT: Solander – (D. & W. 45) S.C. Diptera & Aptera as Oniscus chelipes, in algae off France, Atlantic Ocean; (D. & W. 42) C.S.D. as *O. chelipes* with references to Pallas and Baster. Dryander – Catalogue f.221 as finished in colour, Oniscus chelipes mss —— Ocean, S. Parkinson.

NOTES: the name *Oniscus chelipes* was proposed by Pallas in 1766 and Solander was clearly employing Pallas's name.

239.(**3**:22) *Hepatoxylon trichiuri* (Holten, 1802) Hepatoxylidae

DRAWING: finished pencil, three views; *r.* [ink] 'Fasciola tenacissima/Sydney Parkinson pinx[t] 1769'; *v.* [none]. 235 × 291.

MANUSCRIPT: Solander – (D. & W. 45) S.C. Mollusca 1, f.14 as Fasciola tenacissima, in Squalus glaucus, southern ocean, 11 April 1769; (D. & W. 42) f.423 as above 'intra intestina Squalus glauci'. Dryander – Catalogue f.223 as finished without colour, Fasciola tenuissima mss —— Ocean, S. Parkinson.

NOTES: this larval cestode worm was found in the intestine of the blue shark, *Prionace glauca* (Linnaeus, 1758), which was caught near Osnabrugh Island, now Mururoa, south of the Tuamotu group. The fish was drawn by Parkinson, see numbers 53 and 54 in this catalogue.

240.(**3**:23) *Glaucus atlanticus* Forster, 1800 Glaucidae

DRAWING: finished water-colour; left – enlarged with lettering, right – natural size; *r.* [ink] 'Mimus Volutator/Sydney Parkinson pinx[t] ad vivum 1768'; *v.* 'Oct[r]. 4. 1768/Lat. 11.00 N'. 240 × 220.

MANUSCRIPT: Solander – (D. & W. 45) S.C. Mollusca 1, f.23 as Mimus volutator, Atlantic Ocean 4 October 1768, southern ocean 13 March 1769, 11 April 1770; (D. & W. 42) C.S.D. f.413, same data and 'prope novam Hollandiam Lat 35°30′S' (23 April 1770). Dryander – Catalogue f.225 as finished in colour, Mimus volutator mss —— Ocean, S. Parkinson.

NOTES: this pelagic mollusc was described by J. R. Forster in a communication to J. F. Blumenbach (1800). His specimen was collected in the Atlantic Ocean during the *Resolution* voyage.

241.(**3**:24) Unidentified flatworm Class Turbellaria

DRAWING: finished pencil, left Fig. A – dorsal view, centre Fig. C – dorsal view natural size, right Fig. B – ventral view (both A and B are enlarged); *r.* [ink] 'Doris complanata/A, Animal supra/B, — subtus [A and B bracketed] microscopis auctum/C, — supra magnitudine naturali/[captions to letters on figure]/Sydney Parkinson pinx[t] 1769'; *v.* [ink] 'S. Sea Lat. 29.00. Long. 129:20:/Sept[r] 19. 1769'. 235 × 293.

MANUSCRIPT: Solander – (D. & W. 45) S.C. Mollusca 1, f.26 as Doris complanata, southern ocean 19 September 1769 and 13 April 1770; (D. & W. 42) C.S.D. f.409 Lat 29°10′S, Long 159°20′W (Sept.17. 1769) and Lat. 39°27′S Long. 204°10′W (Apr. 13, 1770). Dryander – Catalogue f.225 as finished in colour, Doris complanata Mss —— Ocean, S. Parkinson.

NOTES: the name Doris complanata proposed by Solander seems never to have been employed by later naturalists. *Doris* was a Linnaean (1758) genus name in Mollusca.

242.(**3**:25) Opisthobranch mollusc — Chromodorididae

DRAWING: unfinished pencil; *r.* [none]; *v.* [pencil] 'The narrow outer edge white the next broad & rich orange then white & so blk & white alternately in the middle the/lower part of the animal the same except that it wants the narrow white edge [;] the feelers & tentacula/Vermillion – the bottom is a violet blk./Doris/ [ink] Endeavours River'. 267 × 371.

MANUSCRIPT: Solander – not recognized in any manuscript. Dryander – Catalogue f.225 as sketch without colours, Doris —— New Caledonia, S. Parkinson.

NOTES: the identification of this opisthobranch is uncertain partly on account of the quality of the illustration; it may belong to the genus *Chromodoris*, and superficially resembles *Chromodoris quadricolor*, an Indo-Pacific species illustrated by Thompson (1976).

243.(**3**:26) Unidentified actinarian — order Actinaria

DRAWING: finished water-colour, left – upper side, centre – side view, right – under side; *r.* [ink] 'Actinia natans/Sydney Parkinson pinx[t] 1770'; *v.* [ink] 'South Sea April y[e] 18th. 1770'. 237 × 295.

MANUSCRIPT: (D. & W. 45) S.C. Mollusca 1, f.44 as Actinia natans, southern ocean, 12 April 1770; (D. & W. 42) C.S.D. f.481, same data, Lat. 39°20′S, Long. 204°8′W. Dryander – Catalogue f.227 as finished in colour, Actinaria natans mss —— Ocean, S. Parkinson.

NOTES: this sea anemone is otherwise unidentifiable; the Solander name Actinaria natans seems not to have been taken up by later naturalists.

244.(**3**:27*a*) *Thalia* sp. — Salpidae

DRAWING: finished pencil, with water-colour, left Fig. I – enlarged with parts lettered, right Fig. II – natural size; *r.* [ink] 'Dagysa gemma/Sydney Parkinson pinx[t] ad vivum. Sept[r] 3rd 1768'; *v.* [ink] 'Sept[r] 2. 1768'. 130 × 240.

MANUSCRIPT: Solander – (D. & W. 45) S.C. Mollusca 1, f.50 as Dagysa gemma on five occasions between 2 September 1768 and 23 April 1770; (D. & W. 42) C.S.D. f.485, same data. Dryander – Catalogue f.229 as finished in colour, Dagysa Gemma mss —— Ocean, S. Parkinson.

NOTES: the genus name *Dagysa*, which Banks and Solander used for all salps was published in Hawkesworth (1773) in his official account of the *Endeavour* voyage. Dagysa gemma, the Solander name, has not been taken up by later authors.

245.(**3**:27*b*) *Pegea* sp. Salpidae

DRAWING: finished pencil and water-colour, three views, one (Fig. III) of a chain of aggregated zooids; *r.* [ink] 'Dagysa saccata/Sydney Parkinson pinx[t] ad vivum Sept[r] 6th 1768'; *v.* [ink] 'Sept[r] 5 1768'. 235 × 259.

MANUSCRIPT: Solander – (D. & W. 45) S.C. Mollusca 1, f.53 as Dagysa saccata, Atlantic Ocean near Spain, 3 September 1768; (D. & W. 42) C.S.D. f.489 as above, August and September 1768. Dryander – Catalogue f.227 as finished in colour, Dagysa saccata mss —— Ocean, S. Parkinson.

NOTES: this Solander name appears not to have been taken up by later naturalists.
Fig. III shows a chain of aggregated zooids probably referable to *Pegea confoederata* (Forsskål, 1775); the two other drawings are of solitary zooids presumed to be individuals of the same species.

246.(**3**:28) *Salpa ?fusiformis* Cuvier, 1804 Salpidae

DRAWING: finished pencil and water-colour; *r.* [ink] 'Dagysa volva/Sydney Parkinson pinx[t] ad vivum 1768'; *v.* [ink] 'Octr 3rd 1768/Lat. 11:11 N.'. 240 × 291.

MANUSCRIPT: Solander – (D. & W. 45) S.C. Mollusca 1, f.55 as Dagysa volva, Atlantic Ocean, 3 October 1768; (D. & W. 42) C.S.D. 491, same data. Dryander – Catalogue f.227 as finished in colour, Dagysa Volva mss —— Ocean, S. Parkinson.

NOTES: the name Dagysa volva seems not to have been taken up by other naturalists. This drawing was copied exactly for Richard Owen's collection of drawings (Owen Colln folio 77). The copy is exact down to the signature and date which caused Ingles & Sawyer (1979) to assume it was a Parkinson original. However, Owen can be exonerated from the suspicion of purloining drawings from the Banks Collection, as the paper on which his drawing is made is watermarked J. Whatman 1805.

247.(**3**:29) *Sulculeolaria* sp. Diphyidae

DRAWING: unfinished pencil, upper enlarged view, lower probably natural size; *r.* [ink] 'Dagysa limpida/Sydney Parkinson pinx[t] 1768'; *v.* [ink] 'Oct[r] 4th 1768/Lat. 09:00N'. 239 × 294.

MANUSCRIPT: Solander – (D. & W. 45) S.C. Mollusca 1, f.57 as Dagysa limpida, Atlantic Ocean, 4 October 1768; (D. & W. 42) C.S.D. f.493, same data. Dryander – Catalogue f.229 as finished without colour, Dagysa limpida mss —— Ocean, S. Parkinson.

Notes: this drawing shows very little detail but apparently represents the posterior nectophore of this siphonophore. The name Dagysa limpida appears never to have been used by later naturalists.

248.(**3**:30) *Cyclosalpa pinnata* (Forsskål, 1775) Salpidae

Drawing: finished pencil and water-colour; four drawings; *r.* [ink] 'Dagysa lobata/Sydney Parkinson pinx^{t} ad vivum Sept 7th 1768'; *v.* [ink] 'Sep^{tr} 6th 1768'. 240 × 294.

Manuscript: Solander – (D. & W. 45) S.C. Mollusca 1, f.58 as Dagysa lobata, Atlantic Ocean, 4 September 1768; (D. & W. 42) C.S.D. f.495, same data. Dryander – Catalogue f.227 as finished in colour, Dagysa lobata mss —— Ocean, S. Parkinson.

Notes: the four drawings represent a whorl of aggregated zooids (Fig. 4), two views of a separate aggregated zooid (with luminous organs picked out in colour and labelled 'g') (Figs. 1 & 2), and a natural size specimen (Fig. 3).
The name Dagysa lobata has not been used by later naturalists.

249.(**3**:31*a*) *Thalia* sp. Salpidae

Drawing: finished pencil; four views; *r.* [ink] 'Dagysa cornuta/[pencil] Sydney Parkinson pinx^{t} ad vivum Septr 2nd 1768'; *v.* [ink] 'Septr 2nd 1768'. 135 × 235.

Manuscript: Solander – (D. & W. 45) S.C. Mollusca 1, f.60 as Dagysa cornuta, Atlantic Ocean, 2 September & 6 September 1768, 6 October 1769; (D. & W. 42) C.S.D. f.497 same dates and 11 April and 23 April 1770. Dryander – Catalogue f.229 as finished without colour, Dagysa cornuta mss —— Ocean, S. Parkinson.

Notes: the name Dagysa cornuta does not seem to have been employed by later naturalists.

250.(**3**:31*b*) *Chelophyes* sp. Diphyidae

Drawing: finished pencil and water-colour; side view; *r.* [ink] 'Dagysa vitrea/ Sydney Parkinson pinx^{t} 1768'; *v.* [ink] 'Octr 7th 1768/[indecipherable]'. 230 × 255.

Manuscript: Solander – (D. & W. 45) S.C. Mollusca 1, f.62 as Dagysa vitrea, Atlantic Ocean 7 October 1768, southern ocean 3 February 1769 & 13 April 1770, f.64, name only; (D. & W. 42) C.S.D. f.499, same data. Dryander – Catalogue f.227 as finished in colours, Dagysa vitrea mss —— Ocean, S. Parkinson.

Notes: the name Dagysa vitrea seems not to have been employed by later naturalists. This appears to be a finished drawing, lettered for captions, by its date made on the first occasion the animal was captured; later drawings are listed under 251 in this catalogue.

251.(**3**:32) *Chelophyes* sp. Diphyidae

Drawing: finished pencil, one with water-colour; *r.* [ink] 'Dagysa vitrea/Sydney

Parkinson pinx[t] 1769'; *v.* [ink] 'March 3[d] 1769. Lat 36.49'/Long. 111.30' W.'. 238 × 295.

MANUSCRIPT: Solander – see above, no.250. Dryander – Catalogue f.227 as finished without colour, —— Ocean, S. Parkinson.

NOTES: these four drawings of the siphonophore were made in the southern Pacific, but the date does not coincide with any of the dates given by Solander (see above), unless 3 March 1769 was written in error for 3 February.

252.(**3**:33) *Thetys vagina* Tilesius, 1802 — Salpidae

DRAWING: finished pencil and water-colour; *r.* [ink] 'Dagysa rostrata./Sydney Parkinson pinx[t] ad vivum Sept[r] 8th 1768/T.13 P.8 Sept. 6. 1768'; *v.* [ink] 'Sept[r] 6th 1768'. 234 × 292.

MANUSCRIPT: Solander – (D. & W. 45) S.C. Mollusca 1, f.66 as Dagysa rostrata, Atlantic Ocean 1768, southern ocean 2 October 1769; (D. & W. 42) C.S.D. f.503, as above 'Mari Atlantico Hispaniam alluente prope fretum herculis . . ., in Oceano Australis, Lat 37°10'S, Long 171°5'W . . .'. Dryander – Catalogue f.227 as finished in colours, Dagysa rostrata mss —— Ocean, S. Parkinson.

NOTES: Solander's name Dagysa rostrata appears not to have been used by later naturalists. This beautifully executed drawing is fully captioned by letters to relate to the description. The annotation T.13 P.8 shows it to have been drawn on the first leg of the voyage, when other Parkinson drawings were similarly labelled. The 'fretum herculis' of Solander's description refers to the pillars of Hercules, said to have stood at the western entrance to the Mediterranean.

253.(**3**:34) *Thetys vagina* Tilesius, 1802 — Salpidae

DRAWING: finished pencil and water-colour; *r.* [ink] 'Dagysa strumosa./Sydney Parkinson pinx[t] ad vivum Sept[r] 8th 1768'; *v.* [none]. 230 × 294.

MANUSCRIPT: Solander – (D. & W. 45) S.C. Mollusca 1, f.68 as Dagysa strumosa, Atlantic ocean near Straits of Gibraltar, off New Holland 23 April 1770; (D. & W. 42) C.S.D. f.505 as above, New Holland = 35°36'S. Dryander – Catalogue f.227 as finished in colour, Dagysa strumosa mss —— Ocean, S. Parkinson.

NOTES: the Solander name Dagysa strumosa appears not to have been employed by any later naturalist. It is surprising that Solander should have recognized two nominal species from specimens in the same area and only two days apart.

254.(**3**:35) *Iasis zonaria* — Salpidae

DRAWING: finished pencil; three views; *r.* [ink] 'Dagysa serena/Sydney Parkinson pinx[t] 1769'; *v.* [pencil] 'Dagysa serena/South Sea Oct[r] 2 1769'. 238 × 280.

MANUSCRIPT: Solander – (D. & W. 45) S.C. Mollusca 1, f.70 as Dagysa serena, southern ocean 2 October 1769, 11 January 1770; (D. & W. 42) C.S.D. f.507, same data, latitude and longitude given. Dryander – Catalogue f.229 as finished without colours, Dagysa serena mss —— Ocean, S. Parkinson.

NOTES: Solander's name Dagysa serena appears not to have been taken up by later naturalists.

255.(**3**:36*a*) *Halistemma* sp. Agalmidae

DRAWING: finished pencil; three views; *r.* [ink] 'Dagysa polyedra/Sydney Parkinson pinx^t 1769'; *v.* [pencil] 'Dagysa polyedra/[indecipherable]'. 200 × 270.

MANUSCRIPT: Solander – (D. & W. 45) S.C. Mollusca 1, f.72 as Dagysa polyedra, southern ocean, 2 October 1769; (D. & W. 42) C.S.D. f.511, same data. Dryander – Catalogue f.229 as finished without colour, Dagysa polyedra mss —— Ocean, S. Parkinson.

NOTES: the name Dagysa polyedra seems not to have been used by later naturalists. Two of the views on this drawing are captioned by letters and the three figures are discussed in Solander's manuscript (D. & W. 42).

256.(**3**:36*b*) Unidentified salp. Order Salpida

DRAWING: unfinished pencil; *r.* [ink] 'Dagysa costata'; *v.* [ink] 'Rio Janeiro'. 164 × 148.

MANUSCRIPT: Solander – not found. Dryander – Catalogue f.229 as finished in colour, Dagysa costata mss —— Rio Janeiro, S. Parkinson.

NOTES: Dryander appears to have been mistaken in claiming this drawing to be finished in colour; there is no water-colour in it. This is the only drawing of Banks and Solander's 'Dagysas' not also to be described; several other Brazilian animals were drawn but not described, presumably because the naturalists were absorbed with their botanical studies at this landfall.

257.(**3**:37) *Physalia physalis* (Linnaeus, 1758) Physaliidae

DRAWING: finished water-colour; *r.* [ink] 'Holothuria Physalis/Sydney Parkinson pinx^t ad vivum.'/*v.* [none]. 371 × 269.

MANUSCRIPT: Solander – (D. & W. 45) S.C. Mollusca 1, f.80 as *Holothuria physalis*, Atlantic Ocean; f.83 surface between the Tropics, 7°S lat; f.84 Atlantic Ocean 22 & 23 December 1768; (D. & W. 42) C.S.D. f.391 and f.393. Dryander – Catalogue f.229, one of two finished in colour, *Holothuria physalis* L. —— Ocean, S. Parkinson.

NOTES: Solander recognized the siphonophore, known as the Portuguese Man-of-War, from Linnaeus's earlier description of *Holothuria physalis* which was based on a number of descriptions and figures given by earlier voyagers including Hans Sloane and Patrick Browne, both travellers to Jamaica, and his own former student Per Osbeck who visited the East Indies in 1750 to 1752. Solander and Banks studied *Physalia* closely and made the earliest observations on the nematocysts, and the manner in which these organisms steer by means of the sail-like membrane. This and subsequent drawings show the membrane in various sailing postures.

Totton (1960) refers to Parkinson's drawings in discussing the history of knowledge of the morphology of *Physalia*, and quotes extensively from Banks's notes on the siphonophore. Moreover he claimed that a specimen in the British Museum (Natural History) register number 1925.8.13.2, preserved in alcohol was a specimen from the *Endeavour* voyage. The label *Holothuria physalis*, was indicative of this, and the handwriting and label design corresponded well with contemporary labels on fishes known to have been captured during the voyage, which I was able to show him. Whether he was correct in claiming it to be the specimen captured on 7 October 1768 south of the Cape Verde Islands, and not one of the later captures, there is no means of knowing.

This drawing was reproduced by Lysaght (1980) at Plate XIVb.

258.(**3**:38) *Physalia physalis* (Linnaeus, 1758) Physaliidae

DRAWING: finished water-colour; *r.* [ink] 'Holothuria Physalis./Sydney Parkinson pinx[t] 1768./[pencil] pepete tata'; *v.* [none]. 371 × 270.

MANUSCRIPT: see above, number 257 in this catalogue.

NOTES: see number 257 above.

259.(**3**:39) *Physalia physalis* (Linnaeus, 1758) Physaliidae

DRAWING: unfinished water-colour; *r.* [none]; *v.* [ink] 'Dec[r] 23 1768/Lat. 37 South./[pencil] N° 4. Holothuria angustata'. 365 × 262.

MANUSCRIPT: Solander – (D. & W. 42) C.S.D. f.395 as Holothuria angustata, habitat in Oceano Atlantico America australis, Lat. aust. gr 37° (December 22, 23, 1768). Dryander – Catalogue f.229 as finished in colours, Holothuria angustata mss —— Ocean, S. Parkinson.

NOTES: Solander apparently considered that the specimens of this siphonophore collected in the South Atlantic were different from Linnaeus's *Holothuria physalis*. His name H. angustata seems never to have been taken up by later naturalists. However, the notes in (D. & W. 45) S.C. Mollusca 1, f.84 refer to *H. physalis* being captured on 22 and 23 December 1768, and these must refer to the specimen drawn. Totton (1965) regarded the genus *Physalia* as monotypic.

260.(**3**:40) *Physalia physalis* (Linnaeus, 1758) Physaliidae

DRAWING: unfinished pencil sketches of six colonies, plus details of tentacles; *r.* [ink] 'S. Parkinson'; *v.* [ink] 'Dec[r] 23. 1768/Lat 37 South/[pencil] N° 4 Holothuria angustata'. 374 × 264.

MANUSCRIPT: see above, no.259 in this catalogue. Dryander – Catalogue f.229 sketch with colours —— Ocean, S. Parkinson.

NOTES: see number 259 in this catalogue.

261.(**3**:41) *Physalia physalis* (Linnaeus, 1758) Physaliidae

DRAWING: pencil with some water-colour of two colonies; *r.* [ink] 'S. Parkinson';

v. [pencil] 'The bladder of this animal is quite transparent/11 Holothuria obtusata/ March 3[d] 1769 Lat. 36.49′ L. 113.3′.' 295 × 236.

MANUSCRIPT: Solander – (D. & W. 45) S.C. Mollusca 1, f.86 as Holothuria obtusa, Pacific Ocean, 3 February 1769, 11 January, 11 & 23 April 1770; (D. & W. 42) C.S.D. f.397 same data, except February 13, 1769 is given, and ... 'an junior H. Physalis Lin. et Mscr?' added. Dryander – Catalogue f.229 as sketch with colours, Holothuria obtusata mss —— Ocean, S. Parkinson.

NOTES: this is the third of the nominal species of *Physalia* which Solander recognized; the name seems not to have been used by later workers.

262.(**3**:42) *Ocyropsis* sp. Ocyropsidae

DRAWING: finished pencil; *r.* [ink] 'Callirrhoe bivia/S. Parkinson pinx[t] 1768'; *v.* [ink] 'Lat. [indecipherable]'. 142 × 236.

MANUSCRIPT: Solander – (D. & W. 45) S.C. Mollusca 1, f.99 as Calliroe bivia surface of tropical Atlantic; (D. & W. 42) C.S.D. f.401 same data, Latitude 7°N. Dryander – Catalogue f.231 as finished without colour, Callirhoe bivia mss —— Ocean, S. Parkinson.

NOTES: the genus name *Callirhoe* was published by Peron & Lesueur (1810) to include two species, *Callirhoe micronema* from the north-east coast of New Holland, and *C. basteriana* from the Dutch coast. They made no reference to Solander's manuscript name and must have derived the name independently. Their observations on the species from New Holland were presumably made during the voyage on the French corvettes *Le Géographie* and *Le Naturaliste* on the expedition led by Baudin. Although C.-A. Lesueur examined the Parkinson drawings and annotated some this was probably after the publication of his (and Peron's) work on medusa, when in 1815 he visited London en route for North America (Goy, 1980).

263.(**3**:43) *Athorybia rosacea* (Forsskål, 1775) Athorybiidae

DRAWING: finished water-colour, two views plus detail of tentacles; *r.* [ink] 'Medusa rutilans/Sydney Parkinson pinx[t] ad vivum 1768./[pencil] voisine lui Lisophisa rosacea. L.S. tableau du radiaire molasses Compoter'; *v.* [ink] 'Octr. 1768/between the tropicks'. 266 × 374.

MANUSCRIPT: Solander – (D. & W. 45) S.C. Mollusca 1, f.102 as Medusa rutilans, Atlantic Ocean between the Tropics; (D. & W. 42) C.S.D. f.445, same data. Dryander – Catalogue f.233 as finished in colour, Medusa rutilans mss —— Ocean, S. Parkinson.

NOTES: the annotation in French refers to Peron & Lesueur's (1810) established family of 'Radiaires molasses compotés', but they did not refer to the name Medusa rutilans. It was written by Lesueur.

This drawing was reproduced by Wheeler (1983) as Plate 186, and by Totton (1954) as Plates II and III.

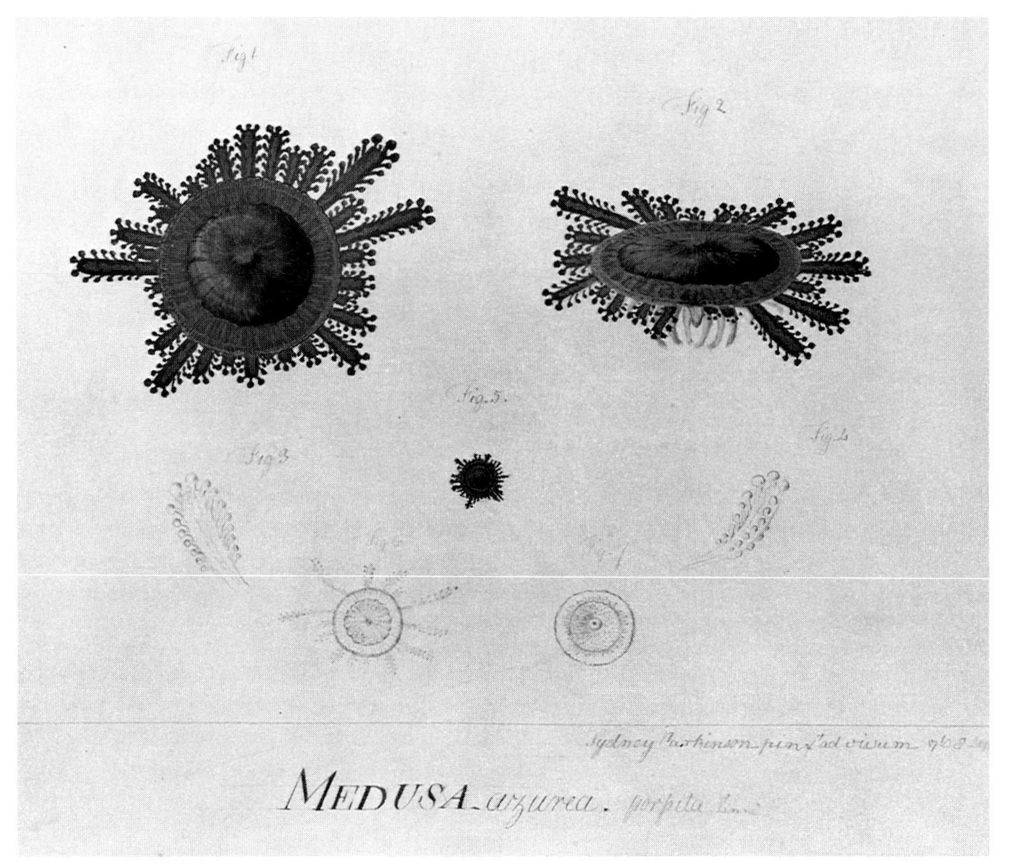

Fig. 18 *Porpita porpita* (Linnaeus, 1758). Parkinson's only drawing on vellum in the *Endeavour* zoological drawings, made between Madeira and the Canary Islands. (Catalogue number 264.)

264.(**3**:44) *Porpita porpita* (Linnaeus, 1758) Velellidae

DRAWING: finished water-colour on vellum, two views enlarged, one view natural size, four studies of detail; *r.* [ink] 'MEDUSA azurea. [pencil] porpita Linné/ Sydney Parkinson pinx^t ad vivum 1768 Sept'; *v.* [none]. 198 × 256.

MANUSCRIPT: Solander – (D. & W. 45) S.C. Mollusca 1, f. 104 as *Medusa porpita*, Atlantic between Madeira and the Canary Islands; southern ocean 13 April 1770; (D. & W. 42) C.S.D. f.447 same data, the Pacific Ocean locality given as Lat. 39°27′S Long 204°10′W. Dryander – Catalogue f.233 as finished in colour, *Medusa porpita* L. —— Ocean, S. Parkinson.

NOTES: Solander's name Medusa azurea does not seem to have been used by later naturalists. However, the reidentification to *M. porpita* Linné must have been made at a relatively early date for it to have been listed by Dryander under this name. This is the only zoological drawing from the *Endeavour* voyage to have been drawn on vellum.

265.(**3**:45) *Porpita porpita* (Linnaeus, 1758) Velellidae

DRAWING: finished pencil drawings, three views; *r.* [ink] 'Medusa porpita./ Sydney Parkinson pinx[t] ad vivum 1768'; *v.* [ink] 'Near the line in the Atlantic'. 240 × 293.

MANUSCRIPT: Solander – see above, no.264 in this catalogue. Dryander – Catalogue f.233 as finished without colour, Medusa Porpita L. —— Ocean, S. Parkinson.

NOTES: see above, no.264 in this catalogue.

266.(**3**:46) *Chrysaora quinquecirrha* (Desor, 1848) Pelagiidae

DRAWING: finished pencil drawing; *r.* [ink] 'Medusa punctulata./Sydney Parkinson pinx[t] ad vivum 1768./[pencil] apartenant au genre Crysaore'; *v.* [ink] 'Brasil'. 371 × 268.

MANUSCRIPT: Solander – (D. & W. 45) S.C. Mollusca 1, f.107 as Medusa punctulata, Rio de Janeiro; (D. & W. 42) C.S.D. f.449, same data. Dryander – Catalogue f.233 as finished without colour, Medusa punctulata mss —— Ocean, S. Parkinson.

NOTES: the Solander name, Medusa punctulata, seems not to have been utilized by other naturalists. The annotation in French (like others in this group of animals) is presumed to be in the hand of C.A. Lesueur, who examined the collection of drawings, probably when he visited England in 1815 on his way to North America (Goy, 1980).

267.(**3**:47) *Dactylometra* sp. Pelagiidae

DRAWING: finished water-colour; *r.* [ink] 'Medusa plicata. Sydney Parkinson pinx[t] 1769'; *v.* [pencil note indecipherable] [pencil] 'No.6 [indecipherable] Becalm'd off Terra del Foego/Lat. 54:23 Jan[ry] 12. 1769'. 240 × 297.

MANUSCRIPT: Solander – (D. & W. 45) S.C. Mollusca 1, f.110 as Medusa plicata, between Tierra del Fuego and Staten Land; (D. & W. 42) C.S.D. f.453, same data. Dryander – Catalogue f.233 as finished in colour, Medusa plicata mss —— Ocean, S. Parkinson.

NOTES: Solander's name, Medusa plicata, seems not to have been used by later naturalists. The unreadable pencil annotation appears to be a note by Lesueur.

268.(**3**:48) *Aequorea* sp. Aequoreidae

DRAWING: pencil and water-colour; four views of the animal; *r.* [ink] 'Medusa radiata/Sydney Parkinson pinx[t] ad vivum 1768'; *v.* [ink] 'off the mouth of the Harbour of Rio de Janeiro'. 238 × 294.

MANUSCRIPT: Solander – (D. & W. 45) S.C. Mollusca 1, f.112 as Medusa radiata, off Rio de Janeiro, in Atlantic, 13 April 1770 (*sic*), and New Holland 23 April 1770; (D. & W. 42) C.S.D. f.455, same data. Dryander – Catalogue f.233, as finished in colours, Medusa radiata mss —— Ocean, S. Parkinson.

NOTES: Solander's name, Medusa radiata, was not used by later naturalists, although the name was independently proposed by Tilesius (1802) from specimens collected at the mouth of the Tagus, Portugal. He made no reference to the Parkinson drawing or Solander's manuscript.

269.(**3**:49) *Aequorea* sp. Aequoreidae

DRAWING: finished water-colour; two views; *r.* [ink] 'Medusa fimbriata./Sydney Parkinson pinx[t] ad vivum 1768'; *v.* [ink] 'off the mouth of the Harbour of Rio de Janeiro'. 268 × 370.

MANUSCRIPT: Solander – (D. & W. 45) S.C. Mollusca 1, f.113 as Medusa fimbriata, Rio de Janeiro harbour; (D. & W. 42) C.S.D. f.459 same data. Dryander – Catalogue f.233 as finished in colours, Medusa fimbriata mss —— Ocean, S. Parkinson.

NOTES: Solander's name Medusa fimbriata has not been used by later naturalists. The name was independently proposed by Dalyell (1848) but there is no evidence that he had seen either Parkinson's drawing or Solander's manuscript in naming what is evidently a different taxon. *Medusa fimbriata* is also listed by Haeckel (1879) attributed to Patrick Browne (1756) but this is not acceptable as a binominal name on account of its date.

270.(**3**:50) *Geryonia proboscidalis* (Forsskål, 1775) Geryonidae

DRAWING: finished pencil; three views; *r.* [ink] 'Medusa chrystalina./Sydney Parkinson pinx ad vivum 1768'; *v.* [ink] 'off the mouth of the Harbour of Rio Janeiro'. 240 × 297.

MANUSCRIPT: Solander (D. & W. 45) S.C. Mollusca 1, f.116 as Medusa crystallina, off Brasil; (D. & W. 42) C.S.D. f.461, same data. Dryander – Catalogue f.233 as finished without colour, Medusa crystallina mss —— Ocean, S. Parkinson.

NOTES: Solander's name Medusa crystallina appears not to have been taken up by any later naturalist.

271.(**3**:51) Unidentifiable species Order Hydroida

DRAWING: finished pencil; two views; *r.* [ink] 'Medusa limpidissima/Sydney Parkinson pinx[t] 1769'; *v.* [pencil] 'N° 7 Medusa limpidissima/Becalmd off terra del Fuego/Lat. 54:28 Jan. 12 1769'. 235 × 295.

MANUSCRIPT: Solander (D. & W. 45) S.C. Mollusca 1, f.117 as Medusa limpidissima, Tierra del Fuego; (D. & W. 42) C.S.D. f.463, same data. Dryander – Catalogue f.233 as finished without colour, Medusa limpidissima mss —— Ocean, S. Parkinson.

NOTES: the Solander name Medusa limpidissima seems not to have been adopted by later naturalists.

272.(**3**:52) *Phialidium* sp. Campanulariidae

DRAWING: finished pencil; four views; *r.* [ink] 'Medusa obliquata./Sydney Parkinson pinx[t] 1769'; *v.* [ink] 'off the Island of terra del Foego/Lat. 54:23. Jan. 12 1769'. 238 × 295.

MANUSCRIPT: Solander – (D. & W. 45) S.C. Mollusca 1, f.119 as Medusa obliquata, near Tierra del Fuego; (D. & W. 42) C.S.D. f.465, habitat America Australis Terram Magellanicum, Lat. 54°23′S; Dryander – Catalogue f.233 as finished without colour, Medusa obliquata mss —— Ocean, S. Parkinson.

NOTES: the Solander name Medusa obliquata seems not to have been used by later naturalists.

273.(**3**:53) *Cyanea* sp. Cyaneidae

DRAWING: finished pencil; *r.* [ink] 'Medusa pellucens./Sydney Parkinson pinx[t] ad vivum 1768'; *v.* 'Oct[r]. 29. 1768/Coast of Brasil'. 374 × 268.

MANUSCRIPT: Solander – (D. & W. 45) S.C. Mollusca 1, f.120 as Medusa pellucens, off Brasil; (D. & W. 42) C.S.D. f.467, Habitat Pelago Oceani Atlantici . . . Brasilia. Dryander – Catalogue f.233 as finished without colours, Medusa pellucens mss —— Ocean, S. Parkinson.

NOTES: the name *Medusa pellucens* was published by Macartney (1810) from a communication from Joseph Banks. Sherborn's *Index Animalium* attributes it to Banks and Solander, but clearly the author of the name was Solander. It is one of the very few names to be published from Solander's extensive work on pelagic Cnidaria.

274.(**3**:54) *Pelagia noctiluca* (Forsskål, 1775) Pelagiidae

DRAWING: finished water-colour; three views of whole animal, two detailed drawings; *r.* [ink] '*Medusa pelagica*. Linn./[pencil] Sydney Parkinson pinx[t] ad vivum 1768/[pencil] voisin de la meduse Panopyre del'atlas du voiage aux terre australia/plan. XXI au doit de la place . . . Crysaone de Peron – a Lesueur'; *v.* [ink] 'Aug[st]. 28. 1768'. 236 × 283.

MANUSCRIPT: Solander – (D. & W. 45) S.C. Mollusca 1, f.122 as *Medusa pelagica*, Atlantic Ocean on several dates, New Zealand 23 April 1770; (D. & W. 42) C.S.D. f.471, localities and dates of capture given. Dryander – Catalogue f.233 as finished in colours, *Medusa pelagica* L. —— Ocean, S. Parkinson.

NOTES: Solander identified the specimen illustrated with *Medusa pelagica* Linnaeus, 1758, and presumably the other specimens given this name by him were in fact referable to *Pelagia noctiluca*. The annotation in French is believed to have been made by C.A. Lesueur, and the reference to planche XXI is to the unpublished plates of Peron (1807) which are referred to by number in Peron & Lesueur (1809). Pl. XXI referred to *Aequorea phosperiphora* from the coast of Arnhemland.

275.(**3**:55) Unidentified — Order Rhizostomeae

DRAWING: unfinished pencil, whole animal and detail; *r.* [ink] 'S. Parkinson'; *v.* [pencil] 'The whole animal a tawny brown (umber & egam) the spots white/the small tentaclae & the triangular ones white./[pencil] Medusa circinnata/Botany Bay'. 368 × 268.

MANUSCRIPT: Solander – (D. & W. 45) S.C. Mollusca 1, f.124 as Medusa circinnata, Sting Rays bay, New Holland; (D. & W. 42) C.S.D. f.469, same data. Dryander – Catalogue f.233 as sketch without colour, Medusa circinata mss —— N.C. (New Caledonia), S. Parkinson.

NOTES: Solander's name Medusa circinnata does not seem to have been used by later naturalists.

276.(**3**:56) *Velella velella* (Linnaeus, 1758) — Velellidae

DRAWING: finished water-colour – two views, water-colour and pencil – two views, pencil – one view; *r.* [ink] 'Phyllodoce velulla/Sydney Parkinson pinx[t] ad vivum 1768'; *v.* [none]. 238 × 293.

MANUSCRIPT: Solander – (D. & W. 45) S.C. Mollusca 1, f.126 as Phyllodoce velella, Atlantic Ocean 7 October 1768, southern ocean several dates; (D. & W. 42) C.S.D. f.475, description and list of captures on six occasions with latitude and longitude given. Dryander – Catalogue f.231 as Phyllodoce vellella mss, finished in colours —— Ocean, S. Parkinson.

NOTES: Solander recognized these solitary floating hydroids (the By-the-wind-Sailor) as the Linnaean species *Medusa velella* but created the genus Phyllodoce to distinguish them from the other cnidarians encountered. It appears not to have been formally published.

277.(**3**:57) *Velella velella* (Linnaeus, 1758) — Velellidae

DRAWING: unfinished pencil sketches; four views; *r.* [ink] 'Phyllodoce velella'; *v.* [pencil] 'N.1. the middle inclosure wt the tentacula white w[t] a pale cast of blue the tentacula at the edge/deep blue – the thin membrane fine blue very deep toward the edge & full of small dots./N.2 the sail quite hyaline the middle inclosures of a dirty blue mark'd w[t] a bright/blue the outer membrane very deep blue especially at the edge'. 235 × 288.

MANUSCRIPT: Solander – see above. no.276 in this catalogue. Dryander – Catalogue no entry for sketch without colours.

NOTES: these appear to be preliminary sketches for the finished drawing listed above, see no.276.

278.(**3**:58*a*) *Beroe* sp. — Beroidae

DRAWING: finished pencil and water-colour; *r.* [ink] 'Beroe marsupium'; *v.* [none]. 126 × 238.

MANUSCRIPT: Solander – (D. & W. 45) S.C. Mollusca 1, f.134 as Beroe marsupium, Atlantic Ocean; (D. & W. 42) C.S.D. f.435 habitat in Oceano Atlanticis intra tropicus . . . var. of B. bilabiata MSS. Dryander – Catalogue f.235 as finished in colours, Beroe marsupium mss —— Ocean, S. Parkinson.

NOTES: Solander's name Beroe marsupium seems not to have been employed by later naturalists.

279.(**3**:58*b*) *Beroe* sp. Beroidae

DRAWING: finished water-colours, eight views, one pencil detail; *r.* [ink] 'Beroe labiata./Sydney Parkinson pinx[t] ad vivum'; *v.* [ink] 'in the harbour of Rio de Janeiro'. 238 × 265.

MANUSCRIPT: Solander – no entries under B. labiata, entries under Beroe bilabiata at (D. & W. 45) S.C. Mollusca 1, f.135, Atlantic Ocean, and (D. & W. 42) C.S.D. f.431, habitat Atlantiis intra tropicos Lat. N. Sinu Oceani Atlantici ad Janeiram Brasilia. Dryander – Catalogue f.235 as finished in colours, Beroe labiata mss —— Rio Janeiro, S. Parkinson.

NOTES: the difference in name on the drawing (Beroe labiata) and in Solander's manuscripts (B. bilabiata) suggests either that Parkinson was responsible for a *lapsus calami* in labelling his drawing, or that Solander changed the name after the drawing was made. The former seems more probable.

280.(**3**:59) *Beroe* sp. Beroidae

DRAWING: finished water-colour; *r.* [ink] 'Beroe incrassata./Sydney Parkinson pinx[t] 1769'; *v.* [ink] 'Becalmd off Terra del Foego/Lat. 54:23 Jan[r.] 12 1769/ [indecipherable pencil note]'. 296 × 235.

MANUSCRIPT: Solander – (D. & W. 45) S.C. Mollusca 1, f.137 as Beroe incrassata, Atlantic near Tierra del Fuego; (D. & W. 42) C.S.D. f.437 same data, locality also given for Oceano Australi on October 2, 1769. Dryander – Catalogue f.235 as finished in colour Beroe incrassata mss —— Ocean, S. Parkinson.

NOTES: Solander's name Beroe incrassata does not appear to have been adopted by later authors.

281.(**3**:60*a*) *Beroe* sp. Beroidae

DRAWING: finished water-colours, one natural size, one enlarged; *r.* [ink] 'Beroe corollata/Sydney Parkinson pinx[t] ad vivum 1768'; *v.* [pencil] 'Straw Colour/ [pencil – inked over] off the mouth of the/harbour Rio Janeiro'. 134 × 234.

MANUSCRIPT: Solander – (D. & W. 45) S.C. Mollusca 1, f.139 as Beroe corolata, Atlantic near Brasil; (D. & W. 42) C.S.D. f.439 same data. Dryander – Catalogue f.235 as finished in colours, Beroe corollata mss —— Rio Janeiro, S. Parkinson.

NOTES: the name Beroe corolata used by Solander was not taken up by later naturalists.

282.(**3**:60*b*) *Hormiphora* sp. Pleurobrachiidae

DRAWING: finished pencil and water-colour; one natural size, two enlarged; *r.* [ink] 'Beroe coarctata./Sydney Parkinson pinx^t 1769'; *v.* [ink] 'South Sea Octr. 2. 1769'. 235 × 269.

MANUSCRIPT: Solander – (D. & W. 45) S.C. Mollusca 1, f. 140 as Beroe coarctata, southern ocean 2 and 6 October, 1769; (D. & W. 42) C.S.D. f.433, localities given as Lat. 37°10′S, Long 171°5′W – 2 October 1769, and Lat. 39°12′S, Long. 174°W – 6 October 1769. Dryander – Catalogue f.235 as finished in colour; Beroe coarctata mss —— Ocean, S. Parkinson.

NOTES: Solander's name Beroe coarctata appears not to have been employed by later naturalists.

283.(**3**:61) *Callianira* sp. Callianiridae

DRAWING: finished pencil; *r.* [ink] 'Beroe biloba/Sydney Parkinson pinx^t. 1770'; *v.* [ink] 'South Sea April ye 13th. 1770'. 296 × 234.

MANUSCRIPT: Solander – (D. & W. 45) S.C. Mollusca 1, f. 142 as Beroe biloba, southern ocean 13 April 1770; (D. & W. 42) C.S.D. f.441 same data, locality Lat. 39°27′S, Long. 204°10′W. Dryander – Catalogue f.235 as finished without colour, Beroe biloba mss —— Ocean, S. Parkinson.

NOTES: Solander's name Beroe biloba appears not to have been taken up by later naturalists.

284.(**3**:62) *Tropiometra carinata* (Lamarck, 1816) Tropiometridae

DRAWING: unfinished pencil; *r.* [ink] 'S. Parkinson'; *v.* [ink] 'Rio Janeiro/ [pencil] N°2 Asterias radiata'. 372 × 265.

MANUSCRIPT: Solander – not found; Dryander – Catalogue f.237 as sketch without colour, Asterias radiata mss —— Rio Janeiro, S. Parkinson.

NOTES: Solander appears not to have described this featherstar, unless he did so for the Slip Catalogue and the sheets were lost early on. However, a number of Brazilian animals appear to have been illustrated but not described. The name Asterias radiata seems not to have been employed by later naturalists.

285.(**3**:63) *Tropiometra carinata* (Lamarck, 1816) Tropiometridae

DRAWING: unfinished pencil and water-colour; *r.* [ink] 'S. Parkinson'; *v.* [ink] 'Rio Janeiro'/[pencil] No 2 Asterias radiata'. 372 × 264.

MANUSCRIPT: see above, no.284 in this catalogue. Dryander – Catalogue f.237 as sketch with colours, Asterias radiata mss —— Rio Janeiro, S. Parkinson.

NOTES: see above, no.284 in this catalogue.

286.(**3**:64) *Culcita novaeguineae* Müller & Troschel, 1842 Oreasteridae

DRAWING: unfinished pencil; *r.* [ink] 'S. Parkinson'; *v.* [pencil] 'Asterias crassisoma/[ink] Otahite'. 372 × 270.

MANUSCRIPT: Solander – (D. & W. 40c) P.A.O.P. f.121 (241) as Asterias crasissima; Dryander – Catalogue f.237 as sketch without colour, Asterias crasissima mss —— Society Islands, S. Parkinson.

NOTES: the name Asterias crasissima seems not to have been adopted by later naturalists.

287.(**3**:65) *Culcita novaeguineae* Müller & Troschel, 1842 Oreasteridae

DRAWING: unfinished water-colour with pencil outline; *r.* [ink] 'S. Parkinson'; *v.* [pencil] 'Asterias crassisoma/[ink] Otahite'. 369 × 271.

MANUSCRIPT: Solander – see above, no.286 in this catalogue. Dryander – Catalogue f.237 as sketch with colours, Asterias crasissima mss —— Society Islands, S. Parkinson.

NOTES: see above, no.286 in this catalogue.

288.(**3**:66) Unidentified starfish Order Valvata

DRAWING: unfinished pencil; *r.* [ink] 'S. Parkinson/[pencil] upper side convex/under side a little concave./the tentacula in the openings reddish orange./Tootooreà'; *v.* [pencil] 'Asterias crassissima/[ink] Otahite'. 375 × 270.

MANUSCRIPT: Solander – see above, no.286 in this catalogue. Dryander – Catalogue f.237 as sketch without colours, see above, no.286.

NOTES: see above, n.286 in this catalogue. It is clear that there is an error in the labelling, probably of this drawing, as Solander's Asterias crassissimus refers to the sea urchin figured in drawing no.286.

289.(**3**:67) *Conchoderma auritum* (Linnaeus, 1767) Lepadidae

DRAWING: finished water-colour, two views; *r.* [ink] 'Lepas Midas./Sydney Parkinson ad vivum pinx[t] 1768'; *v.* [pencil] 'dark purple brown [ink] Nov[r]. 1768/on the bottom of our/ship between the tropicks'. 239 × 291.

MANUSCRIPT: Solander – not found. Dryander – Catalogue f.241 as finished in colours, Lepas Midas mss —— Ocean, S. Parkinson.

NOTES: although Solander described other lepadomorph cirripedes, including Lepas vittata from off the bottom of the *Endeavour* in the Atlantic he apparently made no reference to Lepas midas. The name seems not to have been used by later workers.

290.(**3**:68*a*) *Conchoderma virgatum* var. *hunteri* Darwin, 1851 Lepadidae

DRAWING: finished pencil, two views; *r.* [ink] 'Lepas pelluscens./Sydney

Parkinson pinx^{t} ad vivum 1768'; *v.* [ink] 'Octr. 29. 1768'. 188 × 265.

MANUSCRIPT: Solander – (D. & W. 45) S.C. Mollusca 2, f.86, as Lepas pelluscens surface off Brasil; (D. & W. 42) C.S.D. f. 383 as above, habitat in Pelago Brasiliano, *Medusa pellucenti* adnatus. Dryander – Catalogue f.241 as finished without colour, Lepas pelluscens mss —— Ocean, S. Parkinson.

NOTES: Solander's name Lepas pelluscens seems not to have come into general use. The identification of this drawing should be regarded as tentative.

291.(**3**:68*b*) *Conchoderma virgatum* (Spengler, 1790) Lepadidae

DRAWING: finished water-colour by A. Buchan; *r.* [ink] 'Lepas vittata/ [pencil] A. Buchan Pinx^{t} 1768'; *v.* [ink] 'Novr. 1768/[pencil] Lepas vittata/ [ink] on the bottom of our ship'. 170 × 264.

MANUSCRIPT: Solander – (D. & W. 45) S.C. Mollusca 2, f.88 as Lepas vittata, on the *Endeavour* between the Canaries and Brasil; (D. & W. 42) C.S.D., same data. Dryander – Catalogue f.241 as finished in colour, Lepas vittata mss —— Ocean, A. Buchan.

NOTES: Solander's name Lepas vittatus was not listed by Darwin (1851) but according to Sherborn (*Index Animalium*) was used; he attributed it to Solander (1786) in the *Catalogue* of the Portland collection.

292.(**3**:69) *Lima lima* (Linnaeus, 1758) Limidae

DRAWING: finished water-colour; *r.* [pencil] 'Ostrea Limanda/Sydney Parkinson pinx^{t} ad vivum 1768'; *v.* [ink] 'Brasil'. 236 × 288.

MANUSCRIPT: Solander – not found. Dryander – Catalogue f.241 as finished in colours, Ostrea limanda mss —— Brasil, S. Parkinson.

NOTES: it appears that no description of Ostrea limanda was made by Solander. This is yet another example of an animal from Brazilian waters which was drawn but not described.

293.(**3**:70) *Lima lima* (Linnaeus, 1758) Limidae

DRAWING: unfinished water-colour; two views; *r.* [ink] 'Ostrea limanda Linn.'; *v.* [pencil] 'Ostrea/limanda B.'. 237 × 286.

MANUSCRIPT: Solander – see above, no.292 in this catalogue. Dryander – Catalogue f.241.

NOTES: as Dryander listed only the one drawing and that as finished in colours, this drawing would seem not to have been listed. The attribution of *Ostrea limanda* to Linnaeus is probably a *lapsus calami* for *Ostrea lima* Linnaeus, 1758. See also no.292 in this catalogue.

294.(**3**:71*a*) *Janthina janthina* (Linnaeus, 1758) Janthinidae

DRAWING: finished water-colour by A. Buchan; two views; *r.* [ink] 'Helix violacea/ [pencil] Buchan del'; *v.* [ink] 'Octr. 7th. 1768/[pencil] Helix violacea'. 180 × 266.

MANUSCRIPT: Solander – (D. & W. 45) S.C. Mollusca 14, f.17 as Helix violacea, Atlantic Ocean between the tropics; (D. & W. 42) C.S.D. f.415 same data, also note on observations on species similar in Oceano Australi. Dryander – Catalogue f.243 as finished in colours, Helix violacea L. —— Ocean, A. Buchan.

NOTES: the name Helix violacea does not appear to have been taken up by later authors. No such species was included by Linnaeus despite Dryander's attribution.

295.(**3**:71*b*) *Janthina janthina* (Linnaeus, 1758) Janthinidae

DRAWING: finished water-colour by A. Buchan; *r.* [ink] 'Helix violacea./[pencil] Buchan del.'; *v.* [ink] 'Octr. 7th. 1768 [this overlying pencil] Helix violacea'. 180 × 266.

MANUSCRIPT: Solander – see above, no.294 in this catalogue. Dryander – Catalogue f.243 as finished in colour. Helix violacea —— Ocean, A. Buchan.

NOTES: see above, no.294 in this catalogue.

296.(**3**:72*a*) *Janthina ?globosa* Swainson, 1822 Janthinidae

DRAWING: finished water-colour by A. Buchan; two views; *r.* [ink] 'Helix Janthina/[pencil] Buchan del'; *v.* [ink] 'Octr. 7th. 1768 [overlying pencil]'. 185 × 203.

MANUSCRIPT: Solander – (D. & W. 45) S.C. Mollusca 14, f.19 as *Helix janthina*; (D. & W. 42) C.S.D. f.417 references to Linnaeus and figure. Dryander – Catalogue f.243 as finished in colour, *Helix janthina* L. —— Ocean, A. Buchan.

NOTES: this purple sea snail was referred by Solander to *Helix janthina* Linnaeus, 1758, then the only species of the group recognized.

297.(**3**:72*b*) *Leptaxis* sp. Helicidae

DRAWING: finished water-colour by A. Buchan; two views; *r.* [pencil] 'Buchan del.'; *v.* [ink] 'Madera'. 178 × 264.

MANUSCRIPT: Solander – not found. Dryander – Catalogue f.243 as finished in colour, Helix —— Madeira, A. Buchan.

NOTES: as this drawing is not named it is impossible to reconcile it with any species named by Solander. However, there are no species of Madeiran *Helix* named in the manuscripts and it appears that Solander did not describe it.

298.(**3**:73) *Scutus breviculus* (Blainville, 1817) Fissurellidae

DRAWING: finished water-colour; two views; *r.* [pencil] 'Patella/[ink] S.

Parkinson del'; *v.* [ink] 'Motuaro'. 267 × 372.

MANUSCRIPT: Solander – not found. Dryander – Catalogue f.245 as finished in colour, *Patella* —— New Zealand, S. Parkinson.

NOTES: there is no trace of a *Patella* from New Zealand in the Solander manuscripts, and Wilkins (1955) found only eight New Zealand shells in the Banksian shell collection none of which was this species.

299.(**3**:74) Unidentified sea cucumber — Class Holothuroidea

DRAWING: unfinished pencil sketch; *r.* [ink] 'Parkinson del'; *v.* [pencil] 'Alcyonium anguillare/ [ink] off the Island of Terra del Foejo/Lat. 54:23. Jan^ry. 12 1769'. 262 × 379.

MANUSCRIPT: Solander – (D. & W. 45) S.C. Mollusca 14, f.62 as Alcyonium anguillare, Atlantic Ocean near Tierra del Fuego; (D. & W. 42) C.S.D. f.479 same data 'serpente marino a nauticis'. Dryander – Catalogue f.249 as sketch without colours, Alcyonium —— Ocean, S. Parkinson.

NOTES: Solander's manuscript name Alcyonium anguillare seems not to have been employed by later naturalists. His reference to the name used by the *Endeavour*'s sailors suggests why he employed the trivial epithet.

ACKNOWLEDGEMENTS

A study such as this, which spans many disciplines of zoology and bibliography, owes much to the help the author has received from colleagues and friends both during the immediate period of work and earlier while it was in its formative stages. I must here record my sincere thanks to many such colleagues and friends for their patient help and expertise placed willingly at my disposal.

The following members of staff of the British Museum (Natural History) assisted by identifying the animals represented in the drawings. In the Department of Zoology Dr E.N. Arnold, Dr G.A. Boxshall, Mr R.A. Bray, Miss B. Brewster, Miss A.M. Clark, Dr P.F.S. Cornelius, Mr G.S. Cowles, Dr J.D. George, Mr J.E. Hill, Mr K.H. Hyatt, Dr R.W. Ingle, Dr R.J. Lincoln, Mr F.C. Naggs, Mr G.I.J. Patterson, Dr J.D. Taylor, Mr F.R. Wanless; Department of Entomology Mr M.C. Day, Mrs J.A. Marshall. Members of the staff of the Department of Library Services have been patient with my continuous demands for rare books and access to the drawings and manuscripts, amongst them especially Mrs A. Datta, Miss A.E. Jackson, and Miss J. Jeffrey of the Zoology Library, and Miss D.M. Norman of the General Library. I am also deeply appreciative of the support and help I have received from Mr A.P. Harvey, Head of the Department of Library Services, and his predecessor Mr M.J. Rowlands, as also to Mr R.E.R. Banks of the same Department for his help and patience.

Dr J.E. Randall materially assisted in the identification of many of the Pacific Ocean fishes represented by drawings, as did Dr N.B. Marshall many years earlier; Dr T.W. Pietsch provided information on the manuscript describing *Endeavour* fishes attributed to Solander in Paris; Dr W. Radford and Dr H.E. Brock communicated much information concerning the *Endeavour* invertebrate collections and their fate; to all I offer my thanks. I have also to thank Dr C. Paulin (New Zealand) and the late Dr J.A. Mahoney (Sydney) for help with identifying 'problem' drawings.

Finally, I acknowledge with a deep sense of gratitude several scholars of the Banksian period and the scientific achievements of the *Endeavour* voyage, notably the late Dr A.M. Lysaght, Mr E.W. Groves, Mr J.B. Marshall, and particularly Mr H.B. Carter and Mrs J.A. Diment for their interest and encouragement in producing this catalogue.

Mrs M.B. Newman typed this complex manuscript with her usual competence and patience; I thank her for this essential help.

REFERENCES

[Agassiz, J.L.R.] 1841. A catalogue of fossil fish in the collections of the Earl of Enniskillen, F.G.S., etc. and Sir Philip Grey Egerton, Bart., F.R.S., etc. *Annals and Magazine of Natural History* **7**: 487–498.

Atyeo, W.T. & Petersen, P.C. 1967. The feather mite genus Laminalloptes (Proctophyllodidae: Alloptinae). *Journal of the Kansas Entomological Society* **40**: 447–458.

Bagnis, R., Mazellier, P., Bennett, J. & Christian, E. 1972. *Fishes of Polynesia.* Papeete, Tahiti (Editions du Pacific) 368 pp.

Bauchot, M.L. 1969. Les poissons de la collection de Broussonet au Muséum National d'Histoire Naturelle de Paris. *Bulletin du Muséum National d'Histoire Naturelle* (Ser. 2) **41**: 125–143.

Beechey, F.W. 1839. Introduction. In *The zoology of Captain Beechey's voyage . . . to the Pacific and Behring's Straits performed on His Majesty's Ship Blossom . . . in the years 1825, 26, 27, and 28.* London (H.G. Bohn): i–viii.

Bloch, M.E. 1787. *Naturgeschichte des ausländischen Fische.* **3**. Berlin 146 pp.

Bloch, M.E. & Schneider, J.G. 1801. *Systema ichthyologiae.* Berolini 584 pp.

Blumenbach, J.F. 1800. *Abbildungen naturhistorischer Gegenstände.* Gottingen (J.C. Dieterich) (no pagination).

Boulenger, G.A. 1895. *Catalogue of the fishes in the British Museum.* Second edition. Vol. **1**. London (British Museum (Natural History)) 394 pp.

Boulger, G.S. 1898. Solander, Daniel Charles. *Dictionary of National Biography* **53**: 212–213.

Bourne, W.R.P. 1959. A new Little Shearwater from the Tubaui Islands: *Puffinus assimilis myrtae* subsp. nov. *Emu* **59**: 212–214.

Broussonet, P.M.A. 1780. Mémoire sur les différentes espèces de chiens de mer. *Histoire de l'Académie Royale des Sciences* **1780**: 641–680.

Broussonet, P.M.A. 1782. *Ichthyologia, sistens piscium descriptiones et icones.* Decas 1. P. Elmsly (London) (no pagination).

Browne, P. 1756. *The civil and natural history of Jamaica.* London (for the author) 503 pp.

Buffon, G.L.L. de, 1779. *Histoire naturelle des Oiseaux.* Tom. 6. Paris (l'Imprimerie Royale) 703 pp.

Burgess, W.E. 1978. *Butterflyfishes of the world: a monograph of the family Chaetodontidae.* Neptune City (T.F.H. Publications Inc) 832 pp.

Carr, D.J. 1983. The identity of Captain Cook's kangaroo. *In* Carr, D.J. (Editor) *Sydney Parkinson, artist of Cook's Endeavour voyage.* London & Canberra (British Museum (Natural History) & Australian National University Press): 242–249.

Carter, H.B., Diment, J.A., Humphries, C.J. & Wheeler, A. 1981. The Banksian natural history collections of the *Endeavour* voyage and their relevance to modern taxonomy. *In* Wheeler, A. & Price, J.H. (Editors) *History in the service of systematics.* London (Society for the Bibliography of Natural History (Special Publication 1)): 61–70.

COLLETTE, B.B. & NAUEN, C.E. 1983. *FAO species catalogue*. Vol. **2** *Scombrids of the World* . . . Rome (F.A.O.) 137 pp.

CUVIER, G. 1829*a*. *In* Cuvier, G. & Valenciennes, A. *Histoire naturelle des Poissons*. Vol. **3**. Paris (G. Levrault) 500 pp.

CUVIER, G. 1829*b*. *In* Cuvier, G. & Valenciennes, A. *Histoire naturelle des Poissons*. Vol. **4**. Paris (Levrault) 518 pp.

CUVIER, G. 1830*a*. *In* Cuvier, G. & Valenciennes, A. *Histoire naturelle des Poissons*. Vol. **5**. Paris (Levrault) 499 pp.

CUVIER, G. 1830*b*. *In* Cuvier, G. & Valenciennes, A. *Histoire naturelle des Poissons*, Vol. **6**. Paris (F.G. Levrault) 560 pp.

CUVIER, G. 1831. *In* Cuvier, G. & Valenciennes, A. *Histoire naturelle des Poissons*. Vol. **7**. Paris (Levrault) 531 pp.

CUVIER, G. 1832. *In* Cuvier, G. & Valenciennes, A. *Histoire naturelle des Poissons*. Vol. **8**. Paris (Levrault) 509 pp.

DALYELL, J.G. 1848. *Rare and remarkable animals of Scotland*. **2** London (J. van Voorst) 322 pp.

DARWIN, C. 1851. *A monograph on the sub-class Cirripedia. The Lepadidae; or, pedunculated cirripedes*. London (Ray Society) 400 pp.

DAWSON, W.R. (Editor) 1958. *The Banks letters. A calendar of the manuscript correspondence of Sir Joseph Banks*. London (British Museum (Natural History)) 965 pp.

DIMENT, J.A., HUMPHRIES, C.J., NEWINGTON, L. & SHAUGHNESSY, E. 1984. Catalogue of the natural history drawings commissioned by Joseph Banks on the *Endeavour* voyage 1768–1771 held in the British Museum (Natural History) Part 1: Botany: Australia. *Bulletin of the British Museum (Natural History)* (Historical Series) **11**: 1–183.

DIMENT, J.A. & WHEELER, A. 1984. Catalogue of the natural history manuscripts and letters by Daniel Solander (1733–1782), or attributed to him, in British collections. *Archives of Natural History* **11**: 457–488.

DINGERKUS, G. & DE FINO, T.C. 1983. A revision of the orectolobiform shark family Hemiscyllidae (Chondrichthyes – Selachii). *Bulletin of the American Museum of Natural History* **176**: 1–94.

DRURY, D. 1770. *Illustrations of natural history*. **1**. London (Benjamin White) 130 pp.

DRYANDER, J. 1796–1800. *Catalogus bibliothecae historico-naturalis Josephi Banks*. 5 Vols, London (W. Bulmer). (Dates of publication of the volumes: **1** – 1798, **2** – 1796, **3** – 1797, **4** – 1799, **5** – 1800.)

EGERTON, J. 1976. *George Stubbs, anatomist and animal painter*. London (Tate Gallery) 64 pp.

ESCHMEYER, W.N. 1965. Western Atlantic scorpionfishes of the genus *Scorpaena*, including four new species. *Bulletin of Marine Science* **15**: 84–164.

ESCHMEYER, W.N. 1969. A systematic review of the scorpionfishes of the Atlantic Ocean (Pisces: Scorpaenidae). *Occasional Papers of the California Academy of Sciences* **79**: 1–130.

FABRICIUS, J.C. 1775. *Systema entomologiae*. Flenburg & Leipzig (Royal Publisher) 832 pp.

FABRICIUS, J.C. 1787. *Mantissa insectorum*. **1**. Hafniae (C.G. Prost) 348 pp.

FABRICIUS, J.C. 1798. *Supplementum entomologiae systematicae.* Hafniae 572 pp.

FORSTER, J.G.A. 1777. *A voyage round the world in H.M.S. Resolution, commanded by Capt. J. Cook, during . . . 1772–5.* London (B. White, J. Robson, P. Elmsly & G. Robinson) 2 Vols. 602 & 607 pp.

FORSTER, J.R. 1781*a*. Historia Aptenodytae. Generis avium orbi Australi proprii. *Commentationes Societatis Regiae Scientiarum Gottingensis* **3** (1780): 121–148.

FORSTER, J.R. 1781*b*. A natural history and description of the tyger-cat of the Cape of Good Hope. *Philosophical Transactions of the Royal Society of London* **71**: 1–5, Tab. 1.

FORSTER, J.R. 1785. Mémoire sur les albatros. *Mémoires de Mathématique et de Physique, présentés à l'Académie Royale de Sciences, Paris* **10**: 563–572.

FORSTER, J.R. 1788. *Enchiridion historiae naturali inserviens.* Halae (Hemmerde Schwetschke) 224 pp.

FORSTER, J.R. 1844. *Descriptiones animalium quae in itinere ad maris australis terras per annos 1772, 1773, et 1774 suscepto.* Edited H. Lichtenstein. Berlin (Officina Academica) 424 pp.

GARRICK, J.A.F. & PAUL, L.J. 1971. Deletion of the Australian rays *Aptychotrema banksii* and *Trygonorhina fasciata* from the New Zealand elasmobranch fauna. *Zoology Publications from Victoria University of Wellington* **56**: 1–3.

GARRICK, J.A.F. & PAUL, L.J. 1974. The taxonomy of New Zealand skates (suborder Rajoidea), with descriptions of three new species. *Journal of the Royal Society of New Zealand* **4**: 345–377.

GMELIN, J.F. 1789. *Systema Naturae.* (ed. 13), Tom 1. Lugduni (J.B. Delamolliere) 1516 pp.

GOLDSMITH, O. 1791. *An history of the earth and animated nature.* (New edition.) London (F. Wingrave). Vol. **4**, 328 pp.

GOY, J. 1980. Les meduses de François Péron et Charles – Alexandre Lesueur (1775–1810 et 1778–1846) révélées par les vélins de Lesueur. *Bulletin Trimestriel de la Société Géologique Normandie et Amis Museum du Havre.* **67**: 63–76, 27 pl.

GRAY, G.R. 1844. *List of the specimens of birds in the collection of the British Museum. Part III. Gallinae, Grallaea, and Anseres.* London (British Museum) 209 pp.

GRAY, G.R. [1845]. Birds. *In* Richardson, J. & Gray, J.E. (Editors) *The zoology of the voyage of H.M.S. Erebus & Terror.* London (E.W. Janson) 20 pp.

GRAY, J.E. 1843*a*. Fauna of New Zealand: mammals. *In* Dieffenbach, E. *Travels in New Zealand.* Vol. **2**. London (J. Murray): 177–185.

GRAY, J.E. 1843*b*. *List of the . . . Mammalia . . . in the . . . British Museum.* London (British Museum) 216 pp.

GROVES, E.W. 1962. Notes on the botanical specimens collected by Banks and Solander on Cook's First Voyage, together with an itinerary of landing localities. *Journal of the Society for the Bibliography of Natural History* **4**: 57–62.

GÜNTHER, A. 1860. *Catalogue of the acanthopterygian fishes in the collection of the British Museum.* Vol. **2**. London (British Museum) 548 pp.

GÜNTHER, A. 1861. *Catalogue of the fishes in the British Museum.* Vol. **3**. London (British Museum) 586 pp.

GÜNTHER, A. 1862. *Catalogue of the fishes in the British Museum.* Vol. **4**. London (British Museum) 534 pp.

GÜNTHER, A. 1864. *Catalogue of the fishes in the British Museum*. Vol. **5**. London (British Museum) 455 pp.

GÜNTHER, A. 1866. *Catalogue of the fishes in the British Museum*. Vol. **6**. London (British Museum) 368 pp.

GÜNTHER, A. 1870. *Catalogue of the fishes in the British Museum*. Vol. **8**. London (British Museum) 549 pp.

GÜNTHER, A. 1874. Notice of some new species of fishes from Morocco. *Annals and Magazine of Natural History* (4) **13**: 230–232.

GURNEY, R. 1946. Notes on stomatopod larvae. *Proceedings of the Zoological Society of London* **116**: 133–175.

HAECKEL, E. 1879. *Das System der Medusen*. Jena (G. Fischer) 672 pp.

HAWKESWORTH, J. 1773. *An account of the voyages undertaken . . . in the southern hemisphere*. Vol. **2**. London (Strahan and Cadell) 410 pp.

HERBST, J.F.W. 1799. *Versuch einer Naturgeschichte der Krabben und Krebse*. Bd. 3 (1). Berlin (G.A. Lange) 65 pp.

HOPE, F.W. 1845. The autobiography of Johan Christian Fabricius. *Transactions of the Entomological Society of London* **4**: i–xvi.

INGLE, R.W. 1980. *British crabs*. London (British Museum (Natural History) & O.U.P.) 222 pp.

INGLES, J.M. & SAWYER, F.C. 1979. A catalogue of the Richard Owen collection of palaeontological and zoological drawings in the British Museum (Natural History). *Bulletin of the British Museum (Natural History)* (Historical Series) **6**: 109–197.

IREDALE, T. 1913. Solander as an ornithologist. *Ibis*: 127–135.

KAUP, J.J. 1856. *Catalogue of apodal Fish in the collection of the British Museum*. London (British Museum) 163 pp.

KUHL, H. 1820. *Beiträge zur Zoologie und vergleichenden Anatomie*. Frankfurt am Main (Hermannschen Buchhandlung) 212 pp.

LACEPÈDE, B.G.E. 1798. *Histoire naturelle des Poissons*. **1**. Paris (Plassan, Imprimeur-Libraire) 532 pp.

LATHAM, J. 1781. *A general synopsis of birds*. Vol. **1** (1). London (Benj. White) 416 pp.

LATHAM, J. 1782. *A general synopsis of birds*. Vol. **1** (2). London (Benj. White) pp. 417–788.

LATHAM, J. 1783. *A general synopsis of birds*. Vol. **2** (1) and (2). London (Leigh & Sotheby) pp. 1–366, 367–808.

LATHAM, J. 1785. *A general synopsis of birds*. Vol. **3** (1) and (2). London (Leigh & Sotheby) pp. 1–328, 329–628.

LATHAM, J. 1787. *Supplement to the general synopsis of birds*. London (Leigh & Sotheby) 298 pp.

LATHAM, J. 1790. *Index ornithologicus sive systema ornithologiae*. 2 Vols. London (Leigh & Sotheby) pp. 1–466, 467–920.

LATHAM, J. 1802. *Supplement II to the general synopsis of birds*. London (Leigh & Sotheby) 376, lxxiv pp.

LAY, G.T. & BENNETT, E.T. 1839. Fishes. In *The zoology of Captain Beechey's voyage . . . in the years 1825, 26, 27, and 28*. London (Henry G. Bohn): 41–75.

LEIS, J.M. 1977. Systematics and zoogeography of the porcupine fishes (*Diodon*, Diodontidae, Tetraodontiformes), with comments on egg and larval development. *Fishery Bulletin* **76**: 535–567.

LESSON, R.P. 1831. *Traité d'ornithologie ou tableau méthodique*. Paris (Levrault) 659 pp.

LINNAEUS, C. 1766–1767. *Systema naturae ... editio duodecima, reformata*. Vol. **1**. Holmiae (L. Salvii) 1327 pp.

LYSAGHT, A. 1957. Captain Cook's Kangaroo. *New Scientist* **1** (14 March 1957): 17–19.

LYSAGHT, A. 1959. Some eighteenth century bird paintings in the library of Sir Joseph Banks (1743–1820). *Bulletin of the British Museum (Natural History)* (Historical Series) **1**: 251–371.

LYSAGHT, A.M. 1980. [Introduction to] *The journal of Joseph Banks*. Vol. **1**. *Guildford (Genesis Publications) 101 pp.*

MACARTNEY, J. 1810. Observations upon luminous animals. *Philosophical Transactions of the Royal Society* **100**: 258–293.

MAHONEY, J.A. & RIDE, W.D.L. 1984. The identity of Captain Cook's quoll *Mustela quoll* Zimmerman, 1783 (Marsupialia: Dasyuridae). *Australian Mammalogy* **7**: 57–62.

MANNING, R.B. 1969. Stomatopod Crustacea of the western Atlantic. *Studies in Tropical Oceanography* **8**: 1–380.

MARCGRAVE, G. 1648. *Historiae rerum naturalium Brasiliae*. *In* Piso, W. *Historia naturalis Brasiliae*. Lugduni Batavorum (F. Hackius) 300 pp.

MARSHALL, J.B. 1977. Daniel Solander. *In* Cook, J. *Journal of H.M.S. 'Endeavour' by Capt. James Cook, 1768–1771*. Facsimile edition. Guildford (Genesis Publications): 40–60.

MARSHALL, J.B. 1978. The handwriting of Joseph Banks, his scientific staff and amanuenses. *Bulletin of the British Museum (Natural History)* (Botany Series) **6**: 1–85.

MATHEWS, G.M. 1925. *The Birds of Australia*. Supplement No. 5: Bibliography of the Birds of Australia Part 2. London (H.F. & G. Witherby): 97–149.

MORRISON-SCOTT, T.C.S. & SAWYER, F.C. 1950. The identity of Captain Cook's kangaroo. *Bulletin of the British Museum (Natural History)* (Zoology Series) **1**: 43–50.

MÜLLER, J. & HENLE, J. 1841. *Systematische beschreibung der Plagiostomen*. Berlin (von Veit) 200 pp. 60 pls.

PALMER, G. 1966. Duplication of folio numbers depicting fishes in Parkinson's unpublished drawings of animals from Cook's first voyage (1768–1771). *Journal of the Society for the Bibliography of Natural History* **4**: 267–268.

PARKER, S.P. (Editor) 1982. *Synopsis and classifications of living organisms*. 2 Vols. New York (McGraw-Hill) 1232 pp.

PARKINSON, S. 1773. *A journal of a voyage to the South Seas, in his Majesty's Ship, the Endeavour*. London (Stanfield Parkinson) 214 pp.

PENNANT, T. 1791. *History of quadrupeds*. London (B. White). 2 vols. 566 pp.

PERON, F. 1807. *Voyage de découvertes aux Terres Australes*. Vol. **1**. Paris (Royal printer) 496 pp.

PERON, F. & LESUEUR, C.A. 1810. Des caractères génériques et spécifiques de toutes les espèces de Méduses connues jusqu'à ce jour. *Annales du Muséum d'Histoire Naturelle (Paris)* **14**: 325–366 (no plates issued).

PETERS, J.L. 1979. *Check-list of birds of the world.* **1**. (Second edition). Cambridge, Mass. (Museum of Comparative Zoology) 547 pp.

RANDALL, J.E. 1955. A revision of the surgeon fish genera *Zebrasoma* and *Paracanthurus*. *Pacific Science* **9**: 396–412.

RANDALL, J.E. 1973. Tahitian fish names and a preliminary checklist of the fishes of the Society Islands. *Occasional Papers of Bernice P. Bishop Museum* **24**: 167–214.

RANDALL, J.E. 1983. A review of the fishes of the subgenus *Goniistius*, genus *Cheilodactylus*, with description of a new species from Easter Island and Rapa. *Occasional Papers of Bernice P. Bishop Museum* **25** (7): 1–24.

RICHARDSON, J. 1842*a*. Contributions to the ichthyology of Australia. *Annals & Magazine of Natural History* **9**: 15–31, 120–131, 207–218, 384–393.

RICHARDSON, J. 1842*b*. Description of Australian Fish [part 1]. *Transactions of the Zoological Society of London* **3**: 69–132, pls. 1–6.

RICHARDSON, J. 1843*a*. List of fish hitherto detected on the coasts of New Zealand. *In* Dieffenbach, E. *Travels in New Zealand.* Vol. **2**. London (J. Murray): 206–228.

RICHARDSON, J. 1843*b*. Report on the present state of the ichthyology of New Zealand. *Report of the British Association for the Advancement of Science* **12** (1842): 12–30.

RICHARDSON, J. 1843*c*. Contributions to the ichthyology of Australia. *Annals and Magazine of Natural History*. **11**: 22–28, 169–182, 352–359, 422–428, 489–498.

RICHARDSON, J. 1844–1845. Ichthyology. *In* Hinds, R.B. *The zoology of the voyage of H.M.S. Sulphur, under the command of Captain Sir Edward Belcher, R.N., C.B., F.R.G.S., etc. during the years 1836–42*. Vol. **1**. London (Smith, Elder & Co.): 53–150, pls. 35–64.

RICHARDSON, J. 1844–48. *Ichthyology of the voyage of H.M.S. Erebus and Terror*. Vol. **2**. London (E.W. Janson) viii, 139 pp. 60 pls.

RICHARDSON, J. 1846. Report on the ichthyology of the Seas of China and Japan. *Report of the British Association for the Advancement of Science* **15** (1845): 187–320.

RICHARDSON, J. 1848. Fishes. *In* Adams, A. *The zoology of the voyage of H.M.S. Samarang; under the command of Captain Sir Edward Belcher, C.B., F.R.A.S., F.G.S. during the years 1843–1846*. London (Reeve, Benham, & Reeve) 28 pp. 10 pls.

ROLFE, W.D.I. 1983. William Hunter (1718–1783) on Irish 'elk' and Stubbs's Moose. *Archives of Natural History* **11**: 263–290.

RONDELET, G. 1554. *Libri de piscibus marinis, in quibus verae piscium effigies expressae sunt*. Lugduni (Matthias Bonhomme). 607 pp.

SHARMAN, G.G. 1970. Observations upon animals made by the naturalists of the "Endeavour". *Queensland Heritage* **2** (2): 3–7.

SHARPE, R.B. 1906. Birds. In *The history of the collections contained in the Natural History Departments of the British Museum*. Vol. **2**. London (British Museum (Natural History): 79–515.

SHAW, G. 1793. *The naturalist's miscellany*. **5**. London (F.P. Nodder) [no pagination].

SHAW, G. 1800. *General zoology, or systematic natural history*. Vol. **1** (2). London (George Kearsley) viii, 249–552 pp.

SLOANE, H. 1707. *A voyage to the Islands Madera, Barbados, Nieves, S. Christophers and Jamaica*. Vol. **1**. London cliv, 264 pp. 160 pls.

SMITH, J.E. 1821. *A selection of the correspondence of Linnaeus and other naturalists, from the original manuscripts.* 2 Vols. London (Longman, Hurst & Co.).

SOLANDER, D. 1786. In *A catalogue of the Portland Museum . . . which will be sold by auction . . . on 24th April, 1786, and the thirty-seven following days* . . . London vii, 194 pp.

STEBBING, T.R.R. 1888. Report on the Amphipoda collected by HMS Challenger during the years 1873–1876. *Reports of the Scientific Results of the Voyage of H.M.S. Challenger* (Zoology) **29**: 1–1737.

THOMPSON, T.E. 1976. *Biology of opisthobranch molluscs.* Vol. **1**. London (Ray Society) 206 pp.

TILESIUS, W.G. 1802. Remerkungen über einige Quallen oder Meergallerten (Medusa Linn.), welche sich im Tagus und an den Portugiesischen Seeufern finden. *Jahrbuch für Naturgeschichte, Leipsiz* **1**: 166–177.

TOTTON, A.K. 1954. Siphonophora of the Indian Ocean together with systematic and biological notes on related specimens from other oceans. *Discovery Reports* **27**: 1–162.

TOTTON, A.K. 1960. Part 1. Natural history and morphology. *In* Totton, A.K. & Mackie, G.O. Studies on *Physalia physalis* (L.). *Discovery Reports* **30**: 301–408.

TOTTON, A.K. 1965. *A synopsis of the Siphonophora.* London (British Museum (Natural History)) 230 pp., 40 pls.

VALENCIENNES, A. 1835. *In* Cuvier, G. & Valenciennes, A. *Histoire naturelle des Poissons,* Vol. **10**. Paris (Levrault) 482 pp.

VALENCIENNES, A. 1839. *In* Cuvier, G. & Valenciennes, A. *Histoire naturelle des Poissons,* Vol. **13**. Paris (Pitois Levrault) 505 pp.

VALENCIENNES, A. 1840. *In* Cuvier, G. & Valenciennes, A. *Histoire naturelle des Poissons,* Vol. **14**. Paris (Pitois Levrault) 464 pp.

VALENCIENNES, A. 1847. *In* Cuvier, G. & Valenciennes, A. *Histoire naturelle des Poissons,* Vol. **19**. Paris (P. Bertrand) 544 pp.

VALENCIENNES, A. 1849. *In* Cuvier, G. & Valenciennes, A. *Histoire naturelle des Poissons,* Vol. **22**. Paris (Bertrand) 532 pp.

WATSON, W. 1779. An account of the blue shark, together with a drawing of the same. *Philosophical Transactions of the Royal Society of London* **68**: 789–790, Tab XII.

WHEELER, A. 1963. The nomenclature of the European fishes of the subfamily Trachinotinae. *Annals and Magazine of Natural History* (13) **5**: 529–540.

WHEELER, A. 1964. Rediscovery of the type specimen of *Forcipiger longirostris* (Broussonet) (Perciformes-Chaetodontidae). *Copeia* **1964**: 165–169.

WHEELER, A. 1981. The Forsters' fishes. *In* Cook, J. *The journal of H.M.S. Resolution 1772–1775.* Facsimile. Guildford (Genesis Publications): 783–801.

WHEELER, A. 1983. Animals. *In* Carr, D.J. (Editor) *Sydney Parkinson: artist of Cook's Endeavour voyage.* London & Canberra (British Museum (Natural History) and Australian National University Press): 195–241.

WHEELER, A. 1984*a*. Daniel Solander – zoologist. *Svenska Linnésällskapets Årsskrift* **1982–83**: 7–30.

WHEELER, A. 1984*b*. Daniel Solander and the zoology of Cook's voyage. *Archives of Natural History* **11**: 505–515.

WHEELER A. 1985. The Linnaean fish collection in the Linnean Society of London. *Zoological Journal of the Linnean Society of London* **84**: 1–76.

WHITEHEAD, P.J.P. 1968. *Forty drawings of fishes made by the artists who accompanied Captain James Cook on his three voyages to the Pacific*. London (British Museum (Natural History)) xxxi pp. 40 pls.

WHITEHEAD, P.J.P. 1969. Zoological specimens from Captain Cook's voyages. *Journal of the Society for the Bibliography of Natural History* **5**: 161–201.

WHITEHEAD, P.J.P. 1978*a*. A guide to the dispersal of zoological material from Captain Cook's voyages. *Pacific Studies* **2**: 51–93.

WHITEHEAD, P.J.P. 1978*b*. The Forster collection of zoological drawings in the British Museum (Natural History). *Bulletin of the British Museum (Natural History)* (Historical Series) **6**: 25–47.

WHITLEY, G.P. 1940. *The fishes of Australia. Part 1 The sharks, rays, devil-fish, and other primitive fishes of Australia and New Zealand*. Sydney (Royal Zoological Society of New South Wales) 280 pp.

WILKINS, G.L. 1955. A catalogue and historical account of the Banks shell collection. *Bulletin of the British Museum (Natural History)* (Historical Series) **1**: 69–119.

WILLUGHBY, F. 1686. *F. Willughbeii ... de historia piscium libri quatuor* ... Oxonii (Theatro Sheldoniano) 343, 30 pp.

ZIMSEN, E. 1964. *The type material of J.C. Fabricius*. Copenhagen (Munksgaard) 656 pp.

SYSTEMATIC INDEX

The order of this listing follows the *Synopsis and classification of living organisms* (Parker, 1982). It includes only the modern nomenclature used in the Catalogue, systematically arranged to family, thence alphabetically, relating the entry to the *Catalogue number*, not page number. A general index follows the Systematic index.

Order Rajiformes
Family Rajidae
Raja nasuta 48
Order Myliobatiformes
Family Dasyatidae
Urolophus testaceus 50
Family Myliobatidae
Myliobatis australis 52
Order Orectolobiformes
Family Hemiscyllidae
Hemiscyllium ocellatum 60
Order Carcharhiniformes
Family Scyliorhinidae
Cephaloscyllium isabella 57
Family Carcharhinidae
Carcharhinus sp. 55, 58, 59
Prionace glauca 53, 54

Class Osteichthyes
Order Anguilliformes
Family Muraenidae
Echidna nebulosa 76
Gymnothorax ocellatus 75
Muraena helena 74
Order Siluriformes
Family Ariidae
Bagre marinus 197
Order Myctophiformes
Family Synodontidae
Saurida gracilis 205
Synodus variegatus 206
Order Gadiformes
Family Moridae
Pseudophycis bachus 78
Order Atheriniformes
Family Exocoetidae
Cypselurus (Poecilocypselurus) poecilopterus 201
Exocoetus volitans 202
Parexocoetus brachypterus 200
Family Belonidae
Platybelone argala 199
Order Beryciformes
Family Holocentridae
Holocentrus ascensionis 149
Sargocentron diadema 152
Sargocentron tiere 159

Family Coryphaenidae
- *Coryphaena hippurus* 80

Family Arripidae
- *Arripis trutta* 155, 156

Family Emmelichthyidae
- *Centracanthus cirrus* 150

Family Lutjanidae
- *Lutjanus fulvus* 209
- *Lutjanus kasmira* 164
- *Lutjanus semicinctus* 151
- *Pterocaesio tile* 160

Family Pomadasyidae
- *Plectorhynchus picus* 162, 167

Family Pentapodidae
- *Monotaxis grandoculis* 148

Family Sparidae
- *Anisotremus surinamensis* 158
- *Archosargus rhomboidalis* 122
- *Diplodus caudimaculata* 139
- *Diplodus sargus* 120
- *Pagellus bogaraveo* 137
- *Pagrosomus auratus* 161

Family Sciaenidae
- *Cynoscion* sp. 168
- *Micropogon ?undulatus* 176

Family Mullidae
- *Upeneichthys porosus* 129
- *Upeneus vittatus* 193

Family Kyphosidae
- *Kyphosus incisor* 115
- *Kyphosus sectatrix* 114

Family Ephippidae
- *Chaetodipterus faber* 116
- *Drepane punctata* 96

Family Chaetodontidae
- *Chaetodon citrinellus* 101
- *Chaetodon lunula* 105
- *Chaetodon trifascialis* 100
- *Chaetodon trifasciatus lunulatus* 102
- *Chaetodon ulietensis* 98
- *Chaetodon unimaculatus unimaculatus* 104
- *Chaetodon vagabundus* 111
- *Heniochus chrysostomus* 112

Family Pomacanthidae
- *Centropyge bispinosus* 210
- *Centropyge flavisimus* 106

Family Gobiidae
Valenciennea strigatus 83
Family Acanthuridae
Acanthurus glaucopareius 110
Acanthurus lineatus 107
Naso lituratus 62
Naso unicornis 103
Zanclus cornutus 108
Zebrasoma scopas 97
Family Gempylidae
Gempylus serpens 183
Rexea solandri 182
Family Scombridae
Acanthocybium solandri 178
Katsuwonus pelamis 187
Sarda sarda 192
Thunnus albacares 191
Order Pleuronectiformes
Family Bothidae
Bothus mancus 95
Bothus podas maderensis 93
Family Soleidae
Gymnachirus nudus 94
Order Tetraodontiformes
Family Balistidae
Aluterus scriptus 68
Balistoides viridescens 65
Cantherhines dumerili 66
Melichthys vidua 64
Rhinecanthus aculeatus 63
Rhinecanthus rectangulus 61
Family Tetraodontidae
Arothron meleagris 67
Arothron stellatus 71
Canthigaster solandri 70
Lagocephalus spadiceus 69
Family Diodontidae
Chilomycterus cf *antillarum* 72
Diodon hystrix 73, 229

Class Reptilia
Order Testudines
Family Cheloniidae
Caretta caretta 45, 46, 47
Chelonia mydas 43, 44

Order Passeriformes
- Family Musicapidae
  - *Oenanthe oenanthe* 42
  - *Turdus falcklandi magellanicus* 38
- Family Motacillidae
  - *Motacilla flava* 41
- Family Emberizidae
  - *Rhamphocelus bresilius* 37
  - *Sporophila caerulescens* 39
  - *Volatina jacarina* 40

Class Mammalia

Order Marsupicarnivora
- Family Dasyuridae
  - *Dasyurus hallucatus* 2

Order Diprotodontia
- Family Macropodidae
  - *Macropus robustus* 5
  - *Macropus* sp. 3, 4

Order Primates
- Family Lorisidae
  - *Nycticebus coucang* 1

Order Artiodactyla
- Family Cervidae
  - *Muntiacus muntjak* 6

INDEX

Index of Animal Names

This index includes only names in the Catalogue of drawings. Names set in italic are validly published taxa; those in roman type are manuscript names quoted from the annotations to the drawings or from manuscripts.

The numbers used in the index refer to the numbered catalogue entry. In only three instances are page numbers given and in these the numeral is preceded by p.

Companion Microfiche Collection

relating to

The Catalogue of the Natural History Drawings

commissioned by

Joseph Banks on the Endeavour Voyage 1768–1771

held in

The British Museum (Natural History)

The materials listed in the *Catalogue* (three parts) have been reproduced in colour and positive image black and white microfiche as described below:

THE MICROFICHE—

1. Botanical Specimens

This microfiche collection is composed of approximately 1,000 specimens, including many types, relating to all of the drawings of plants made on the *Endeavour* voyage.

2. Botanical Drawings and Related Engravings

This microfiche collection contains all of the drawings of plants made by Parkinson on the *Endeavour* voyage. These include detailed watercolour drawings as well as outline drawings in pencil with colour references. Parkinson made over 900 drawings before his death towards the end of the voyage. Banks employed five artists in London to make about 600 finished drawings from Parkinson's outline drawings and these are included in the microfiche. In addition 738 copper plate engravings of the drawings commissioned by Banks and 318 lithographs of the Australian drawings published by James Britten in 1900–1905 are included.

3. Zoological Drawings

This microfiche collection includes all the drawings of animals made by Parkinson (and two other artists) on the *Endeavour* voyage. The collection comprises more than 300 drawings including detailed watercolour drawings as well as outline drawings in pencil.

Available from

Meckler/Chadwyck-Healey Scientific Micropublishing
11 Ferry Lane West, Westport, CT 06880, U.S.A.